An Introduction to Keeping Sheep

An Introduction to Keeping Sheep

JANE UPTON

and

DENNIS SODEN

Farming Press

Acknowledgements

Thanks must go to Gill McLellan, whose cheerful enthusiasm and expert typing of the manuscript was invaluable. Very special thanks go to our long-suffering families for their help and patience over the last long 18 months. Finally, our appreciation goes to all the sheep owners and members of the Small Shepherd's Club who kindly allowed us to use their sheep and properties as valuable source material for our photography and illustrations.

First published 1991

ISBN 0 85236 226 9

A catalogue record for this book is available
from the British Library

Published by Farming Press Books
4 Friars Courtyard, 30-32 Princes Street
Ipswich IP1 1RJ, United Kingdom

Distributed in North America
by Diamond Farm Enterprises,
Box 537, Alexandria Bay, NY 13607, USA

Photographs and drawings by Jane Upton
Cover design by Andrew Thistlethwaite
Phototypeset by Typestylers, Ipswich
Printed and bound in Great Britain by Butler and Tanner, Frome, Somerset

Contents

Foreword

The author and illustrator of this book have performed a great service to the present day sheep industry in collating so much instructive information into one volume.

It therefore gives me much satisfaction to commend this book to sheep people of all ages and creeds. Having been concerned with "the sheep" for most of my life, and having had the good fortune to see it in its various forms in several countries around the world, I find it very obvious that the basic management skills need to be set out in plain language for all to study. Especially including the ever growing "flock" of part-time amateur and novice sheep keepers.

This great sheep industry that spans the United Kingdom today has its roots way back in history. The Jacob Sheep Society claims a mention in Genesis XXX, where Jacob displays his better knowledge in selecting the more prolific breeders. He obviously knew a thing or two, even if, in my opinion, he dishonestly pulls a fast one on his brother.

Then again in the *Odyssey*, Homer describes how the Cyclops managed his milking sheep with their long fleeces, and how Odysseus used his craftiness to escape from the cave hanging on underneath them.

In this country in the Middle Ages, the history of how the monastic houses managed their vast flocks — and contributed to the wealth of the nation — makes fascinating reading.

Even today in the European Community the structure and marketing of sheep products is an integral part of the new European environment. Throughout the ages all these achievements are — and have been — only possible with good management.

This comprehensively compiled book is full of detail, description and a common sense approach to a very wide and diverse subject. However, the novice should not be daunted by the apparently large collection of pitfalls and health hazards portrayed. It really is not all bad, but it is right to be aware of the many possible difficulties.

It has often been said that a sheep will always find a new way to die, and one's boyhood memories of the traditional Dorset winter lambing yard are still very vivid — not all life and joy! But has it ever been considered what the losses could be if a human maternity ward were to have 500 "mums" all giving birth in eighteen days — with the majority being multiples — and only one doctor and two nurses in attendance!

No, if you really pause and consider her, our sheep is a hardy, adaptable and sensible creature. That calm capable look which she gives you in the field, with a healthy lamb standing either side of her, is very reassuring. I never cease to wonder, with enormous pleasure, at every live birth, when that small wet being coughs and takes its first breath.

In conclusion, a memory. A neighbouring farmer of my father's generation, who was a very tough and successful Dorset man, used to say that when he felt depressed, tired or just plain fed up, he would go up to the top end of the farm away from the cows and the cereal crops, and just walk about among his flock of Dorset Down ewes. His equilibrium and peace of mind would then return.

This book will teach you how to achieve that state.

GEOFFREY HYDE

Wool, Dorset
January 1991

Introduction

Those of us who have watched a group of lambs playing in the spring sunshine or seen the miracle of a new-born lamb taking its first breath will readily agree that there is a great deal of pleasure to be experienced in keeping and caring for a flock of sheep.

This book is written from a background of many years of practical experience in large and small flocks of both pedigree and commercial sheep, and it seeks to help and guide those who are keeping sheep for the first time. It describes in detail how and why various practical tasks are carried out and how to avoid some of the pitfalls and awkward situations that can arise from time to time. It describes some of the differences between different types of sheep, and it suggests how, when and where to set about starting a flock.

CHAPTER 1

Getting Started

What has prompted your decision to keep sheep? Are they to act purely as grass mowers with the added bonus of a homegrown, maybe organically produced, supply of meat? Does the prospect of keeping breeding animals appeal? Is there a need to diversify in an existing livestock enterprise?

In any event the sheep will need an area of land on which they can graze. That land will have to be securely fenced, with a supply of water nearby. The prospective owner will have decided whether sufficient spare time is available to care for the sheep throughout the year, and whether help will be available for periods when holidays or sickleave are taken. The amount of time required to give adequate care to the sheep will vary from two short visits per day in the summer after shearing, to frequent visits during each twenty-four hour period during lambing time.

Once these basic requirements have been considered and the decision has been taken to go ahead, it is time to start making plans.

Making contacts

Sheep keepers come from all walks of life, and there might be a local club where people who keep sheep as a hobby can get together and discuss experiences and organise talks, visits and demonstrations. Your local vet should be able to tell you of others who keep small flocks. Visit local sheep fairs and auction sales and meet and talk with as many sheep keepers as possible.

Assessing the available land

The size of your flock will to a large extent be dictated by the amount of land that is available for grazing. Assess the quality and quantity of the herbage growing on it, remembering that you will expect your sheep to live almost entirely from it for up to eight months of the year. Is it 'light' or 'heavy' land? Heavy land is close-textured, usually clay-based and is generally more productive than light land, but it will not stand heavy stocking during periods of high rainfall as it poaches or gets trampled into mud very quickly. Light land is composed of larger particles of soil, is free draining and does not poach so easily, but the herbage growing on it does suffer during periods of hot weather and drought. Remember also when you are assessing the grazing that your flock may increase by as much as two hundred per cent following a very successful lambing.

How many sheep?

Initially stocking rates should be no more than five ewes per hectare. This number can be adjusted in subsequent years if necessary. The sheep will be healthier and more contented if your land is understocked rather than overstocked.

Where to find your sheep

If there is a sheep or shepherd's club operating locally this can be a good source

A sheep auction

as members often have surplus stock for sale. Look at the small ads in your local paper, or your vet may know of sheep for sale. Be wary of small groups being offered for sale in the local auction market. They may be rejects from a large flock for any number of reasons including the ability to escape from any type of enclosure! However, a trip to a local auction market or sheep fair can be useful as a guide to the value of different classes of sheep and it will give you an idea of how much you must expect to pay when you make your purchase.

If you decide that you would like to keep a pure breed of sheep you will be able to obtain the name and address of the secretary of the breed society from the National Sheep Association, The Sheep Centre, Malvern, Worcestershire WR13 6HP. The secretary will be able to let you know where you can obtain specimens of your preferred breed, and whether any are available in your locality. If one of the rare breeds has taken your fancy, details can be obtained from the Rare Breeds Survival Trust, National Agricultural Centre, Stoneleigh, Warwickshire CV8 2LG.

CHAPTER 2 **What Type of Sheep to Buy**

The type of sheep to buy will be determined to a certain extent by what is available and the time of year when you decide to purchase. Most of the sheep fairs and sales of breeding sheep take place between August and November, but sheep from private sources are available at all times of the year. Ewes of four or five years of age that have already produced lambs are probably the best buy for beginners as they will be experienced in motherhood. Another possibility is to buy weaned ewe lambs in the autumn but not to mate them until the autumn of the following year. This will give the beginner an opportunity to get to know the sheep and how to handle them. Ewe lambs are much lighter in weight and the various tasks are more easily carried out with them than with heavier sheep. You will become more proficient in handling as their weight increases!

If the sheep are required purely to keep grass under control, then consider buying wether lambs, i.e. castrated males, as their feed requirements are less than for breeding ewes. They are usually sold by their breeders in the autumn and fattened during the next few months before being sold for meat. If they are kept as grass mowers then they will need to be shorn and dipped in the spring and will have increased considerably in size and weight by the following autumn. They can then be sold for meat and replacement lambs purchased. The main drawback to this is that they will be offered for sale at a time when smaller, younger lambs are also being sold for meat, and the price per kilo tends to decrease as the weight and age of the sheep increase. Yet another possibility is to buy

wether lambs of a longwoolled or finewoolled breed and keep them for a number of years, using the income from their high-value fleeces as a contribution towards the cost of keeping them.

Conformation

There are certain basic standards which should be looked for:
● The sheep should have a good mouth with incisor teeth which meet the pad on the top jaw in the correct position — neither overshot nor undershot.

NORMAL JAW

UNDERSHOT JAW

OVERSHOT JAW

GOOD OVERALL CONFORMATION

(FRONT)

STRONG BONE & STRAIGHT LEGS

- There should be good strong bone and straight legs, not bowing inwards or outwards. Pay particular attention to the hind legs of rams. Ensure that there are no weaknesses in the pastern, which is the joint immediately above the foot. Any weakness there can affect the performance of rams and ewes kept for breeding.
- A long level back is desirable with a good spring of rib giving plenty of room for the heart and lungs.

(BACK)

LEGS SHOULD NOT BE BOWING OUTWARDS OR INWARDS

A WEAK PASTERN

A GOOD FOOT

In addition to the above characteristics choose sheep that have that undefinable quality of 'character' or 'presence' which makes them stand out from other members of the flock.

If selecting a pure breed, see that the breed standards for size, colour and fleece quality are met.

WHICH BREED TO BUY

There are many different breeds of sheep in the United Kingdom and they can be categorised into three main types.

The Shortwools

These include the Down breeds which were developed in the southern half of England by using a Southdown ram on various local breed ewes. The Southdown was fixed as a breed and greatly improved for meat production by selection from within the breed by John Ellman at Glynde near Lewes in Sussex in the eighteenth century. It has been used in the production of such diverse shortwools as the very finewoolled Dorset Down, the Hampshire Down, the very popular Suffolk and the very large Oxford Down. All these breeds were developed in the areas that their name indicates to suit local requirements. Because of their docility and close flocking instincts they were also used to improve the fertility of the soil through a system known as 'folding'. In this system the sheep were confined by a temporary fence of hurdles onto a small area of a crop grown specially for them. Each day they were moved on to a fresh area and confined there so that their dung was distributed evenly over the field. This benefited subsequent grain crops and at the time was one of the principal sources of fertility, especially on outlying areas of the farms.

The Down breeds are now almost entirely used to produce terminal sires for crossing with the Masham, Mule, Scottish

A SHORTWOOL

Halfbred and similar crossbred ewes where the progeny are destined for the meat market. These ewes come from the upland areas and when brought to lowland farms produce quality prime lamb for export and for home consumption. The ability of the Down breed rams as terminal sires has been recognised worldwide and they have been exported to many different countries.

The Dorset Horn and its sister breed the Polled Dorset which was developed in Australia share an important characteristic in that they will breed at almost any time of the year. In some cases it is possible to breed two crops of lambs in one year, but more generally flocks are managed so that they produce three crops of lambs in two years. The ewes are exceptionally good milkers which explains why their progeny can make high quality carcases at ten weeks of age.

The Ryeland   A white-faced breed rather similar in appearance to the Southdown. Originating in the county of Hereford it has been used during its eight hundred year history to improve and develop other breeds. It is a compact and docile breed producing good quality wool and meat.

The Wiltshire Horn   Unique among British sheep, the Wiltshire Horn is one breed that does not require shearing. Shepherding

6

costs are lower in that it rarely gets cast and flystrike is virtually unknown. It is an excellent prime lamb sire and rams are extremely active and prepotent in that they transmit to their lambs liveliness at birth, early maturity and the ability to fatten easily off grass alone.

The Shropshire This medium-sized black-faced breed has lost popularity gradually since its heyday around the turn of the century. It is an alert, upstanding breed with a good conformation, capable of siring prime lamb of good quality under practically all climatic conditions.

Kerry Hill This strikingly attractive breed with its black and white face markings originates from the Kerry Hills in Montgomeryshire on the borders of Wales. The ewes are outstanding milkers and they make first-class mothers of prime lambs when crossed with Down breed sires.

Clun Forest Another breed from the border counties of Wales with a reputation of prolificacy, its popularity has waned during the past twenty-five years, although it is still popular in its home area. It produces prime early maturing lambs, especially when crossed with the blockier Down breeds.

The Longwools

Generally speaking, the longwools are larger sheep which mature more slowly than the shortwools. They were more popular when families were larger and consequently larger joints of mutton were required. Their fleeces provided an income from the relatively poorer areas of hill land of the Cotswolds and the Yorkshire and Lincolnshire Wolds. These areas are the homes of the Cotswold, Lincoln Longwool and Leicester breeds, but the group also includes the Romney Marsh which originates from the low-lying areas of Kent. The Romney has the ability to withstand very high stocking densities and can carry a higher worm burden than other breeds without showing ill-effect. It has been exported successfully to many parts of the world onto all kinds of terrain, and when crossed with the Southdown in New Zealand it was responsible for producing the famous Canterbury lamb. It is still the most popular breed in the area which bears its name.

Some of the longwools, notably the Border Leicester which originates from the Leicester and Cheviot breeds, the Teeswater, Wensleydale and Hexham or Blue-faced Leicester, are used almost exclusively to produce rams for crossing with ewes of the hill breeds. The resulting crossbred ewe lambs are sold to lowland farmers, either as lambs or as eighteen month old ewes, where they are mated with shortwool rams to produce prime lamb for the meat market, the longwool sires having contributed hybrid vigour, prolificacy and early maturity to their crossbred progeny.

Other members of this group include the Dartmoor, the Devon Closewool and the Devon and Cornwall Longwool which, as their names imply, come from the high-rainfall areas of south-west England. Mature rams of the Devon and Cornwall breed can produce around sixteen kilos of wool annually, as can rams of the Lincoln Longwool and Leicester breeds.

A LONGWOOL

The Devon Closewool North Devon, where this breed originates from, and from where it has spread over large areas of south-west England, is one of the most densely sheep-populated areas of the world. Rarely seen outside the south-west, the Devon Closewool is medium-sized with a white face and produces good fleece of three to four kilos of medium-quality wool used for hosiery and tweeds. Many flocks live entirely off grass all the year round.

The Dartmoor The Dartmoor is a breed of sheep whose fleece and constitution have been developed to cope with the extreme climatic conditions of the bleak moorland which is its home. Numerically it is not as strong as many other breeds. The wool is lustrous and fleeces from yearling rams average twelve kilos in weight with a staple or individual fibre length of 225mm.

The Hill Breeds

The hill breeds occupy a unique position in the sheep industry in that their distribution covers a huge area of high land which is largely unsuitable for any other type of farming. Two qualities which all the hill breeds possess are hardiness and thriftiness. Hardiness gives them the ability to withstand severe climatic conditions, and thriftiness the ability to thrive on a limited food supply, often of poor quality.

They are smaller sheep than either of the previous two categories, and, unlike the Down breeds, they have evolved a habit of scattered grazing. They spread themselves very thinly on the ground so that there is less competition for the available grazing. When enclosed in a lowland situation, they regard a field boundary as a challenge to see what lies beyond as they believe the grass is always greener on the other side! This behaviour is understandable when it is appreciated that they are the descendants of

A HILL BREED

animals which have been given free range over a huge area of countryside for many hundreds of years.

The hill breeds include:

The Scottish Blackface, which is very numerous not only in Scotland but also on the moors in south-west England, carries a fleece of long, coarse wool which enables it to endure exposure to the heavy rainfall and gale force winds common to these areas.

The Swaledale is probably the strongest breed of sheep numerically and is found mainly in the north of England. It too is very hardy and thrifty and can withstand the severest weather.

The Dalesbred is a cousin of the Swaledale but with a softer fleece free from black or grey fibres.

The Herdwick holds pride of place for hardiness and can exist on the sparse and poor quality grazing of the high fells of the Lake District under wintry conditions of extreme hardship. It is unlike its neighbours in that it has a white face and a fleece colour that can vary between white, grey, brownish red and black.

*The **Rough Fell** is another sheep with long, coarse wool used for carpets and mattresses which enables the animal to withstand the hardships of its surroundings in north Lancashire and the fells.

*The **Cheviot** takes its name from the Cheviot Hills on the border between England and Scotland. It is more at home on the greener hills than the mountain crags or heather moors, but is still extremely hardy. It is also maintained as a pure breed on the mountains in Wales.

*The **Welsh Mountain** comes in several different types. It varies between the small, extremely hardy sheep of the high mountains and the larger type of the common land around the valleys of South Wales. All have white faces with tan markings and all produce lean carcases with unsurpassed flavour. Any kind of fence used to enclose them will be closely inspected and before long an escape route either through, under or over will have been found!

There are other breeds which represent transitional types between the three main categories. These include the Beulah Speckleface, Llanwenog and Radnor from Wales, the Lonk and Derbyshire Gritstone from Lancashire and the Peak District, and the Exmoor Horn from north Devon. All are bred to suit local conditions and are maintained as pure breeds in the areas in which they are found.

One breed which has now left the ranks of the rare breeds is the Jacob sheep. It is very popular with small flock owners as it is a prolific, commercially viable breed and can be instantly recognised by its black and white patterned fleece. It comes in polled, two-horned or multihorned varieties and its

A RARE BREED: MANX LOGHTAN

devotees claim a high level of resistance to foot problems. The coloured fleece finds a ready market with home spinners, producing wool which knits into attractively shaded patterns.

In recent years many breeds of sheep have been imported from Europe and some are taking an increasingly important role both as prime lamb sires and as sires of crossbred lambs from hill ewes, in competition with the traditional British sheep. Of these the Texel and Bleu de Maine are the most prominent, together with the Charollais and Rouge de l'Ouest, and other important importations like the finewoolled Merino and Polwarth have come from Australia. All of these, together with the indigenous rare breeds which have been protected by the Rare Breeds Survival Trust, give the beginner an enormous selection of breeds from which to choose when starting a flock.

CHAPTER 3 **Planning Your Sheep Area**

FENCING

Before purchasing your sheep, make sure that the enclosure is securely fenced. If the sheep are allowed to escape they can be a constant source of annoyance to yourself and to your neighbours. The fence can be a natural hedge but unless it is one that has recently been laid the sheep will browse on it and eventually create weak spots. Avoid poisonous species such as yew, rhododendron etc. to which they might be attracted at certain times of the year, particularly when grass is in short supply. Brambles can become entangled in the wool so that the sheep is unable to free itself, and the bark can be stripped from hedgerow and cultivated species in the spring when the sap is rising, thus ring-barking and killing trees and bushes.

Far superior is well-erected fence consisting of a stock fence with a code number of C8/80/15. This is of rectangular mesh wire, with the mesh graduated so that the smallest mesh is near the ground so as to prevent the sheep putting their heads through and either getting stuck or exerting undue pressure so as to weaken the fence. Avoid using any stock fencing with a larger mesh size than C8/80/15.

This netting is 900mm high and additional height can be obtained by adding strands of plain or barbed wire above it. A strand of barbed wire below the fence at ground level will prevent stock from creeping underneath and will also discourage dogs and predators coming in from the opposite direction.

Any wire fence is only as sound as its

Stock fencing

STOCK FENCE C8/80/15
WITH ADDED PLAIN WIRE
FOR ADDITIONAL HEIGHT AND
BARBED WIRE AT GROUND LEVEL

HINGED MESH

construction, and it is vitally important that the fence is correctly tensioned and the supporting stakes are not more than two metres apart. The strainer posts at either end of a straight run of fencing should be at least 900mm into the ground, tightly rammed and well strutted. Somewhere on the fence, either at a strainer post or anywhere along its length, fit a means of tightening the wire when it becomes slack through age or frequent atmospheric temperature changes. Eye bolts, winders and radisseurs can be used as tensioners. When erecting the fence do not drive the holding staples into the intermediate stakes beyond the point where they just touch the wire. This will allow the wire to move through the staples when tensioning or when expansion or contraction occurs. The staples should be at a slight angle when driven into the stakes so that they enter different grains in the wood and avoid splitting.

If the fence is to reinforce a hedge, it should be sited a few feet away from the hedge. This keeps the sheep from climbing onto the fence to feed on the growing tips of the hedge plants and subsequently weakening it. Such behaviour can be discouraged by fitting a single line of electric fence wire so that it projects at right angles

A strainer post

TWO TYPES OF STRAINER POSTS WITH STRUTS

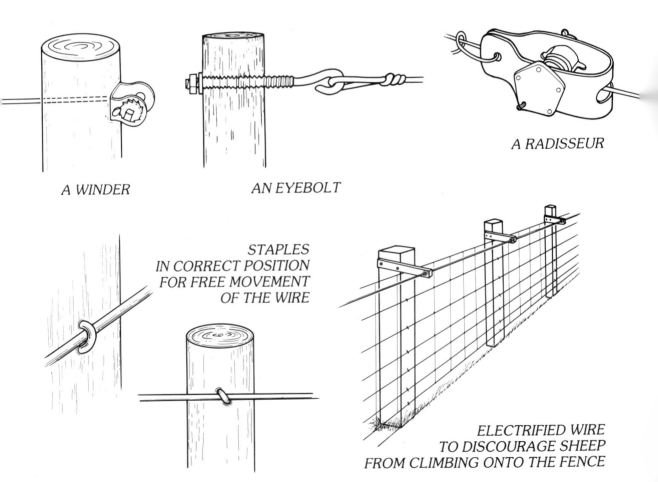

A WINDER

AN EYEBOLT

A RADISSEUR

STAPLES
IN CORRECT POSITION
FOR FREE MOVEMENT
OF THE WIRE

*ELECTRIFIED WIRE
TO DISCOURAGE SHEEP
FROM CLIMBING ONTO THE FENCE*

ELECTRIFIED
NETTING

ELECTRIC FENCE
WITH THREE
LINE WIRES

becoming ineffective through earthing by allowing herbage to grow on to it, or by allowing it to come into contact with metal which is not insulated. Very young lambs do not respect electric fencing in the same way as adult sheep. As their small hooves make very light contact with the ground they receive no shock from the fence, and they can become entangled in the electrified netting. However, electric fencing is a good option when the grazing is rented rather than owned, as it can be taken away if use of the land is lost, and put up somewhere else at no further expense. Portable electric fencing is also very useful for making temporary divisions in a field.

If possible site any gate in the boundary fence near a corner of the grazing area. This will make it easier when moving the sheep. It will also be an advantage if the gate opens through 180 degrees; some of the sheep, especially young lambs, will inevitably go to the wrong side if it is at 90 degrees. If the entrance to the grazing is to be used frequently it is advisable to raise the level of the gateway a few inches above the surrounding land with hard core or scalpings so that it remains free from mud even in the ʹworst weather conditions.

from the intermediate fence posts about 150mm from the top of the wire netting.

Electric fencing can be used as an alternative to the stock fencing. It can be in the form of netting or single line wires. The netting consists of nylon mesh with fine lightweight wire twisted into it to carry the electric current. Single strands of the same materials are also available for use with insulators on wooden posts or on metal posts with built-in insulators. More permanent electric fences using No. 12 galvanised steel wire with insulators on metal or wooden posts does not suffer damage from rabbits and hares as the nylon netting sometimes does. The number of line wires (two or three) and the distance between the wires and the total height of the fence will vary according to the breed.

It is advisable to use an electric fence unit connected via a transformer to the mains electricity supply, or if a portable source of electricity is required use a battery operated fencer unit. Batteries have a habit of running down at critical times, weekends, holidays, etc. The best way to make sure a battery does not run out is to have two, one in the field and the other at home being topped up by a trickle charger. Care should be taken at all times to prevent the electric fencing from

A BATTERY-OPERATED
ELECTRIC FENCER UNIT

A SHEEP HOUSE

HOUSING

The grass on the grazing area makes little or no growth through the winter months and both sheep and land will benefit from being rested. During this time the sheep will need supplementary feed and it makes it more comfortable for sheep and shepherd if the flock can be kept under cover for a few weeks. This is also a convenient way of buying extra acres if the food consumed at this time has been grown elsewhere.

The sheep house need not be an elaborate building, and it is sufficient to use a simple pole barn construction. Posts which have been pressure-treated with preservative e.g. secondhand telegraph poles, will last much longer in the ground than untreated ones. A mono-pitch roof will be sufficient and the building should be sited with its back to the prevailing wind with a good overhang to the roof both front and back. If the front overhang is angled downwards, it will prevent rain from blowing onto the bedding.

Barley straw provides the best bedding material. It is superior to wheat, rye or oat straw, all of which very soon get trampled flat whereas barley straw stays light and open textured for much longer. The straw should be clean, bright and free from mould. The ewes will pick it over, eating most of the 'flag' or leaf from the stems. Soiled bedding should be cleared from the house periodically to prevent the build-up of moist heat which can lead to foot problems.

Allow four square metres for each ewe. This will also allow you to accommodate the lambs if necessary. The building should have solid walls to the back and ends to a height of 1200mm, with Yorkshire cladding, consisting of narrow vertical boards with a 20mm gap between them, above. Heavyweight perforated plastic can also be used but this is difficult to tension. Such construction will allow adequate ventilation with no draught at floor level.

The fence at the front can be combined with a feed barrier which will enable the sheep to be fed without entering the pen. There must be sufficient space for all the sheep including shy feeders to feed comfortably. A minimum width of 500mm per animal is recommended. The entrance to the house should be a gate not less than 2m in width. The ideal floor is made from rammed chalk, but hardcore covered with smaller material will also be satisfactory. The height of the finished floor should be above the level of the surrounding land so as to avoid problems with drainage. Solid concrete floors should be avoided as the bedding quickly becomes damp and respiratory problems can arise. You may prefer to feed your sheep on a hardstanding in front of the building. Again, the feed can be placed in the troughs without interference from the sheep and you will be able to inspect them more easily.

Water should be available at all times in small valve-operated troughs with provision

14

made for any water resulting from leakages or blocked valves being directed outside the building. If there is a supply of water available from a conventional cattle drinking trough, provide a step or build up the surrounding area so that the sheep can drink comfortably. When young lambs are present, float a heavy plank of wood on the surface of the water. This will carry the weight of any lamb jumping onto the trough when playing and save it from drowning. It will also prevent the growth of algae in the trough during warm weather.

HANDLING PENS

When the building is no longer required for holding sheep, it can also be used as a handling pen. First clear all the soiled litter from the floor and give a liberal dressing of lime. For small numbers of sheep the only other equipment needed is two or three hurdles hinged together with string and fixed at one end to a side wall of the house. These can be opened out straight and then closed around the sheep, holding them in a confined space while dosing, vaccinating, etc.

If no housing is available then a small, securely fenced enclosure should be constructed in a corner of the field, preferably near the entrance and exit gate. It should consist of a pen large enough to hold all your sheep together, with a small area for individual handling. It is also advisable to build a funnel approach to the pen as this will make penning so much easier. A piece of hessian or similar material stretched between two people can be useful when driving sheep into a narrowing entrance to the pen. If it is found necessary to have a hard floor to the outdoor handling pen, this can be made from loose pebbles around 25mm to 50mm in diameter. These will revolve as the sheep walk on them and are self-cleaning. A retaining board around the outside of the pen will be required to prevent the pebbles from spreading.

The gates to the outdoor handling pen should be strongly constructed and hung so that they close onto the sheep as they are penned. The fastenings should be about 500mm from the ground as this is where almost all of the pressure exerted by a pen

A sheep house doubling as a handling pen

USING HESSIAN TO DRIVE SHEEP INTO THE PEN

filled with sheep will come. It is a great help to be able to step over the internal fences of the outdoor handling pen and a suggested height for these is 900mm. They should be of solid construction with occasional 'footholes' built in to assist those who are unable to step over. The perimeter fences need to be a little higher at 1,200mm in order to contain the more athletic breeds.

Two things to bear in mind are that sheep will move uphill more freely than downhill, and towards light rather than towards dark, solid, high walls or buildings.

You may also consider it worthwhile having a shallow tray or footbath where sheep requiring treatment for footrot or scald can be stood in a solution of zinc sulphate for a while.

If the sheep are kept on rented land where it is not possible to build a permanent handling area, a temporary one can be made using metal hurdles, securely held together. These are easily moved, and can be taken to wherever the sheep are, so that there is always an area in which they can be caught and held for inspection or treatment.

A TEMPORARY HANDLING PEN WITH FUNNEL SHAPED APPROACH

CHAPTER 4 Learning How to Handle Your Sheep

During the course of the year it is necessary to handle your sheep for dosing, vaccinating, foot trimming, etc. As described in the previous chapter, confine them in a small area so that the handler does not have to put on a rodeo show each time one needs to be caught! Do not be tempted to grab the wool as the sheep finds this very painful. In any event there is no wool on a newly shorn sheep.

CATCHING

There are three ways in which a sheep can be caught and restrained:

1. Firmly grasp a back leg just above the hock and lift the leg off the ground.
2. Grasp the loose skin in the flank of the sheep just in front of a back leg and lift the leg from the ground.
3. Catch the lower jaw of the sheep in a cupped hand and lift the head above its natural horizontal position. Place the fingers of the other hand under the tail. It can now

be moved backwards or forwards, but generally it is easier to move a sheep backwards, keeping the head up all the time.

If you wish to examine the teeth or read an ear number, it will be sufficient to back the sheep into a corner while still keeping control of the head, but you may find it easier to do this when the sheep is completely immobilised.

IMMOBILISING

There are several ways of immobilising a sheep, and the following directions apply to right-handed people.

It can be done by sitting the sheep on its hindquarters and holding it between the knees, leaving both hands free for the examination. This will also be the position required when paring overgrown feet or treating footrot. To achieve this, having got the sheep into position with its chin held by a cupped hand, and standing on its left side, move the left hand around the neck so that

CATCHING BY THE HIND LEG

GRASPING LOOSE SKIN IN THE FLANK

with the left hand cupped under its chin and place the right hand on its right hip bone. Move the sheep close to the legs, with the left knee behind its left shoulder and the right knee in front of its left hip, and pull its head round so that it is looking back over its right shoulder. At the same time press firmly downwards with the right hand and move the right leg smartly away from contact with the sheep and it will collapse onto its hindquarter.

With experience this method of immobilising a sheep requires very little effort, but once the sheep is in position it is

CATCHING THE LOWER JAW OF THE SHEEP IN A CUPPED HAND

Moving the sheep backwards or forwards

the head is resting on the wrist and pull the sheep towards you. Place the left knee just behind the sheep's left shoulder and the right knee just in front of the sheep's left hip. Reach over and firmly grasp the loose skin in front of the sheep's right hind leg. Lean backwards and by using the knees as a pivot the sheep will roll onto its side on the knees of the operator. It can be lowered gently onto either its left or its right hindquarter. Avoid trying to make the sheep sit vertically on its tail as it finds this very uncomfortable.

Another method of sitting the sheep onto its hindquarters is to hold the sheep

MOVING THE SHEEP BACKWARDS OR FORWARDS

IMMOBILISING THE SHEEP BY ROLLING IT ON TO THE KNEES OF THE OPERATOR

AN ALTERNATIVE METHOD OF IMMOBILISING THE SHEEP

20

Immobilising

necessary to prevent its feet from coming in contact with the ground or fences etc. because if they do the sheep will kick and struggle as it attempts to regain the standing position.

There are a number of pieces of equipment on the market which can be used to immobilise sheep. The simplest of these is similar to a stretcher which can be placed at an angle against a fence. The animal is moved backwards until its rear end touches the frame and then the head is lifted until eventually the sheep is lying on its back on the stretcher. A rope is tied across the sheep just below its front legs and the operator is then free to use both hands to treat the sheep. (See page 25.)

Other items include a sheep 'cradle' which uses a series of clamps and levers to secure the sheep and to move it into an upside down position. This is helpful for anyone who has a back problem, but, as with the previous item, it has the disadvantage to the manual method in that the sheep has to be caught and moved to the equipment before it can be treated.

For heavily pregnant ewes who need assistance at lambing time, another method of immobilising is described on pages 44-45.

It is possible to restrain sheep simply by passing the head through a loop of rope fixed to the side of the handling pen and placing the left knee just in front of the animal's left hip while pushing it against the side of the pen. This can be useful while dagging, crutching or treating blowfly strike in the tail area.

To examine the teeth, use the right hand to hold up the head and part the lips, using the thumb and index finger of the left hand to expose both the upper fibrous pad and the incisor teeth in the lower jaw. There are no incisor teeth in the upper jaw.

A SHEEP STRETCHER

A MECHANICAL DEVICE FOR IMMOBILISING THE SHEEP

CHAPTER 5

Caring for Your Sheep

Adequate food and water are two essentials to the wellbeing of your sheep. The actual amount consumed will vary according to the type of sheep kept, and also from season to season, but try to maintain good bodily condition at all times. Be prepared to offer supplementary food when the nutritional requirements of the sheep are not being met, because of poor quality or insufficient grazing. The condition required on any particular type of sheep can only be learned by experience, and a visit to a local market with a knowledgeable companion will be useful.

Supplementary Food

The type of supplementary food offered will depend on the time of year when the grass is in short supply and also on the type of sheep. Good quality hay which has been made from grass cut at the early flowering stage, and baled and carted in dry weather, and with no trace of mould, will usually provide for maintenance if fed at two kilos per day. Ewes in the later stages of pregnancy will also require concentrates. Concentrates are highly nutritious foods e.g. grain and various by-products from the edible oil seed crushing mills. These can be purchased in cube or pellet form to a guaranteed protein content from the local feed merchant. Introduce six weeks before lambing at the rate of 100g per day, increasing to 750g per day at lambing and afterwards. The concentrate is best provided as a compound mix of 17 per cent protein. It is important that the compound is one

Feeding hay during the winter months

Adequate feeding space

specially formulated for sheep, as those prepared for other classes of livestock can contain minerals in proportions injurious to sheep. Similarly, sheep that are being fattened for meat or being prepared for show or sale, may require supplementary food.

Besides good quality hay and the concentrates, it may be necessary to provide more bulk in their diet. All kinds of root and cabbage crops are suitable, although one should watch out for the possibility of choking. NEVER feed green potatoes.

Sugar beet pulp is a very useful food. When fed dry the analysis compares with rolled oats. When soaked overnight in water, the extra bulk is beneficial in giving extra substance to animals being fed for show. If mangolds are available, they should be fed to ewes only after lambing. Any secondary shoots which have grown during storage should be trimmed off before feeding to wethers or rams, as they will cause urinary problems.

Supplementary food can be expensive, and it is important that the sheep obtain the maximum benefit from the food offered. All the feed troughs and water containers should be kept clean and free from contamination from faeces, mud, etc. and there should be sufficient trough space to allow all the sheep to feed in comfort. Hay should be fed from racks or hay boxes and kept dry at all times. If racks are used, try to prevent the sheep from standing directly underneath as the hay seeds and chaff will contaminate the fleeces thus reducing the value of the shorn wool, and they may also be responsible for eye problems.

If you offer young lambs supplementary feed in the form of pellets specially formulated for them, this can be done by fencing off an area to which only the lambs have access. Vertical posts which are spaced so that only the lambs can pass between them, together with a horizontal rail 450mm from the ground, will effectively exclude adult sheep.

Provide access to a balanced mineral supplement in a waterproof container at all times. Check with your vet whether there are any particular mineral deficiencies in your area, and if so how to overcome them.

24

ROUTINE TASKS

In addition to the general requirements listed above, there are specific tasks which periodically require attention.

Foot Trimming

Foot trimming becomes necessary when the horny outer surface of the foot grows beyond the soft, rubbery sole of the foot. Sheep that are housed and those kept on damp, heavy land will require more frequent attention than those kept on drier and more abrasive soils. Feet that are not trimmed regularly can become compacted with soil and other debris, eventually causing lameness. To trim the foot, first immobilise the sheep and using straight-bladed secateurs or a sharp knife, carefully pare away the overgrown horn down to the level of the sole of the foot. Sometimes the pointed front end of the foot will have grown forwards and upwards. Be particularly careful when paring in this area as severe bleeding will occur if it is cut back too far.

*USING SECATEURS TO TRIM
OVERGROWN FEET*

Dosing

In order to keep the sheep free from internal parasites, they will need to be dosed periodically with anthelmintics. These can be in the form of liquids, paste or capsules. Some are broad spectrum, which can deal with a number of different kinds of parasites, while others are more specific. It is important that the manufacturer's instructions for dosing are followed correctly, and as dosage depends on weight it will be necessary to weigh representative individuals to ensure that all the sheep receive the recommended amount of the anthelmintic. Give each sheep a temporary crayon mark as it is dosed to avoid giving a double dose to some and missing others completely!

Sheep can be dosed in the standing position. Hold the head in a horizontal position by placing a hand under the lower jaw and insert the instrument containing the anthelmintic into the corner of the mouth and onto the back of the tongue before releasing the dose. Various types of dosing instrument are available. These include:

- A trigger-operated gun which automatically delivers a pre-determined measured dose from a reservoir carried on the operator's back.
- A paste gun which delivers a measured dose.
- A narrow-necked bottle which will require filling with a measured dose between each sheep.
- A large plastic syringe (minus the hypodermic needle!) which is more suitable for young lambs.

If the sheep starts to cough violently, release the head at once to avoid choking.

DOSING WITH A PASTE GUN

AND

WITH A TRIGGER-OPERATED GUN

DOSING A LAMB WITH A SYRINGE

Vaccination

It is possible to control a large number of diseases affecting sheep by preventative vaccines, and if these are given at the correct times and in the correct quantities, the immunity to the diseases can be passed from the ewe to the lamb via the colostrum, giving protection to the lamb for up to ten weeks. Depending on the type, the vaccine can be administered in one of three different ways.

1. For a ***subcutaneous injection***, lift a loose fold of skin with a thumb and forefinger, insert a hypodermic needle into the fold and release the vaccine by depressing the syringe. The preferred site for injections of small quantities of vaccine is high up on the neck, behind the ears. Quite often a small abscess or nodule will form at the injection site, and this can affect the subsequent value of the skin if present anywhere other than near the edge.

2. ***Intramuscular injections*** should be given at a site where there is good muscle cover, e.g. into the hind leg. The needle is pushed straight into the muscle and the syringe contents released slowly.

3. A ***live vaccine*** is applied to an area of bare skin between the front leg and the chest for protection against orf. The site should be clean but not disinfected otherwise the vaccine might be affected, and after application the site is lightly scratched so that the skin is broken to allow penetration of the vaccine. Gloves should be worn when using this vaccine as orf can also affect humans.

A SUBCUTANEOUS INJECTION (ABOVE) AND AN INTRAMUSCULAR INJECTION (BELOW)

Dagging and Crutching

Both of these operations involve removing wool from the tail area of the sheep. In the spring when grass is growing quickly, the faeces are more liquid and can contaminate the wool. A build-up on the sheep can attract the blowfly to lay its eggs from which maggots will develop. Removing the wool helps to prevent this from happening, and it also avoids contamination of the fleece at shearing time.

A pair of sharp, short-bladed (100mm to 150mm) hand shears is used for dagging. The sheep is secured in a standing position by passing its head through a loop of rope attached to the side of the pen. Place a knee in front of the hip of the sheep, leaving both hands free to complete the operation.

Crutching involves taking off the wool from the area around the genitalia of the ewes prior to mating, but probably the main benefit will be at lambing time when the shorter wool makes visual inspection easier, particularly with longwoolled sheep. If wool is removed in front of the udder and from the front of the hind leg, it will give the newborn lamb easier access to the teats. To do this it will be necessary to turn the ewe as when trimming feet unless you possess a sheep cradle.

Dagging or crutching using hand shears

SECURING A SHEEP

THE SECURED SHEEP

1

2

REMOVING THE WOOL
FROM THE TAIL END OF THE SHEEP
USING HAND SHEARS

3

Shearing

In May or June all sheep that are more than one year old will need to be shorn. This is a skilled specialist operation and best left to shearing contractors. It might be possible to take the sheep to a neighbouring sheep farmer when his flock is being shorn, but the arrangements should be made well in advance as the shearers are usually very busy people.

Sheep are much less likely to suffer from stress at shearing if they are penned and kept away from food for at least twelve hours before they are to be shorn. They sit more easily on the shearing board, and the board stays clean and free from faeces. The wool *must* be dry before the sheep are penned under cover. Shorn fleeces which are wet, and contaminated with faeces, soil or vegetable matter will attract a penalty from the British Wool Marketing Board when the wool is sold.

out on a flat clean surface and any part of the fleece contaminated with faeces, mud, etc. removed. The loose belly wool, if clean, is placed on the centre of the fleece and the sides are folded to the middle before rolling from the tail end towards the neck. Make a small hole in the fleece and tuck the neck wool into this, and place the whole fleece in the sack provided.

Any skin cuts resulting from shearing will need treatment to assist healing. Keep a watch on any cuts that are being plagued by flies and treat again if necessary. The wound may be in a position where the sheep can strike it with its hind feet in an attempt to keep the flies away. This can soon result in an unpleasant open sore and it may become necessary to use lightweight sacking shaped like a horse rug to cover the body of the sheep. It should be inspected daily and dressed with an insecticidal cream. In extreme cases, putting the sheep into a darkened shed in the daytime will give it some respite from the unwelcome attention of flies.

Routine Inspections

All through the year a routine inspection of the flock is necessary, the frequency of the inspections varying with the time of year and the breed and condition of the sheep. Ideally it should never be less than twice a day, and more in the period preceding lambing. At this time the ewes are in good condition and heavy with lambs, and they can get on to their back when rolling and be unable to regain the standing position. This is common in early spring when showery weather alternating with drying sunshine and winds makes the sheep itchy. They roll onto their backs to relieve the itch and then get stuck in that position. If left like this, the continuing fermentation of food in the stomach and the pressure of the unborn lambs combine to restrict the breathing severely, eventually leading to suffocation and death.

Some months before shearing it is necessary to register the flock with the British Wool Marketing Board, Registration Department, Oak Mills, Clayton, Bradford, West Yorkshire BD14 6JD. They will undertake to purchase the fleeces and will provide sacks to pack them in. Their agents will arrange collection after being notified that shearing has been completed. The British Wool Marketing Board requires that the fleeces are rolled in the correct manner. With shortwools the skinside of the fleece is on the outside, and with longwools the skin side is on the inside. The shorn fleece is laid

In the period from early May until early November the sheep maggot fly is active and a careful watch is necessary for any sheep that are 'struck'. These will require prompt treatment so that any injury to the sheep caused by the maggots burrowing into the skin is minimised. A few weeks' protection from the maggot fly is obtained when ewes are newly shorn and also when the lambs have been dipped in an insecticidal dip.

Spend time watching the sheep without disturbing them so that any abnormal behaviour can be quickly recognised and if necessary acted upon.

Teeth

There are no incisor teeth in the sheep's upper jaw! Eight small incisor teeth appear soon after birth in the lower jaw. At about fifteen months of age the centre two are shed, and replaced by two larger, permanent incisors. At yearly intervals a tooth on either side of these will be shed and replaced by permanent ones, and the sheep is known consecutively as a two-tooth, four-tooth or six-tooth, according to the number of permanent incisor teeth present. At just over four years of age an adult sheep should have eight permanent incisor teeth and is then known as full-mouthed.

After four years the teeth continue to grow and will eventually be shed. The sheep is then said to be broken-mouthed. This may lead to feeding problems and unless special provision can be made any such ewes should be culled before mating.

Occasionally sheep are born with a lower jaw shorter or longer than normal. This can lead to feeding difficulties as the teeth will not meet the upper jaw pad correctly, and is an indesirable characteristic. Sheep with this condition are said to be 'undershot' or 'overshot' and should not be used for breeding. Careful examination of the teeth of any sheep which you consider buying is necessary.

THE INCISOR TEETH OF A YOUNG LAMB

OF A 15 MONTH OLD SHEEP

OF A 3 YEAR OLD SHEEP

Any problems with the molars, or back teeth, in either upper or lower jaws can usually be felt as hard lumps through the skin along the line of the jaw. Your vet will advise of any treatment necessary.

BODY CONDITION SCORING

Condition scoring is an attempt to standardise how much fat and muscle a sheep is carrying by using the fingers to assess the degree to which the vertical and horizontal processes of the backbone are covered. The point where the assessment is made is in the region of the loin, behind the last rib and over the kidneys. It is scored on a scale of 0 to 5, and half points are also used.

Although the flock may be all of the same breed or cross, there will be considerable variation between individuals at any particular time. Between different breeds the variation will be even greater, and it will prove to be almost impossible to get all the animals into the same condition at the same time even if they are kept under the same conditions!

ASSESSING THE BODY CONDITION AROUND THE DOCK AND OVER THE LOIN

THE POSITION OF THE HANDS ACROSS THE LOIN

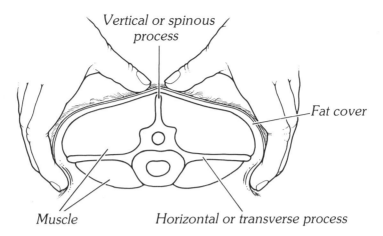

Vertical or spinous process

Fat cover

Muscle

Horizontal or transverse process

Score 0 This indicates extreme emaciation with no fat or muscle detectable.

Score 1 The backbone feels very sharp in both the vertical and transverse processes, and muscle tissue in the angle between is very shallow with no fat cover.

CONDITION SCORE 1 — LEAN

Score 2 The muscle tissue is of sufficient depth to give the processes a smooth and rounded feeling, but there is little fat cover.

Score 3 Individual spinous processes can be felt only by pressure and are smooth and well covered. The loin muscle is full with a moderate fat cover.

CONDITION SCORE 3 — GOOD

Score 4 The loin is full with a thick covering of fat and the spinous processes detectable only as a firm line. The ends of the transverse processes cannot be felt.

Score 5 The animal is very fat. There is a shallow groove in the fat cover over the vertical processes and the transverse processes are undetectable.

CONDITION SCORE 5 — FAT

CHAPTER 6

Caring for the Breeding Flock

CARE OF THE BREEDING EWE

A typical year for the breeding ewe begins a few weeks before mating takes place. The planned size and composition of the flock will have been completed, and the ewes brought onto a rising plane of nutrition, aiming for a body condition score of 3.5 at mating time. The proflicacy of the flock will be increased if the ewes are 'flushed' about three weeks before mating. This involves putting them onto better grazing if it is available, or feeding 200g of concentrates to each ewe each day until mating is completed. Flushing will result in more eggs being shed by the ovaries and more lambs being born. Any parasitic or foot problems will have been corrected, and any young sheep will have been given the appropriate vaccine in order to protect them from the clostridial group of diseases and from pasteurella pneumonia.

When the ram has joined the flock, and mating has begun, all unnecessary stress for the ewes should be avoided. It is best if the ram is fitted with a harness with a marker crayon which puts a colourful mark onto the rear of each ewe as she is mated, as this will

Ewes in good condition

enable you to keep a record of the mating date and work out the expected lambing date for each ewe. The gestation period is about 147 days.

When mating has been completed, remove the ram to a separate enclosure if possible. The condition of the ewes can now be allowed to drop back to a score of 3, although they should be given hay to appetite when the quality of the grazing deteriorates. If housing is available and the weather is bad, consider housing the ewes but try to ensure that the fleeces are dry before they move into the house for any period of time. Wet sheep will take a long time to dry when under cover.

Six weeks before lambing commences, begin concentrate feeding in addition to hay at the rate of 100g per sheep each day. Increase this amount weekly so that each ewe is receiving 750g daily when lambing commences. A booster injection of the vaccine against clostridial diseases should be given two weeks before they are due to lamb.

When each ewe has lambed place her in an individual pen made with hurdles and measuring 1.8m square. Any size smaller than this will expose the newborn lamb to the risk of being lain on when the ewe lies down, especially with multiple births. She should stay in the pen with her lambs for a few days until it is apparent that all is well with both ewe and lambs. The afterbirth will have come away shortly after lambing, and it should be collected and disposed of away from predators. The udder of the ewe will need to be inspected to ensure there is an adequate supply of milk for the lambs, and the lambs castrated and docked using rubber rings. When turning the ewes out of their individual pens with their lambs to join others that have already lambed, it is advisable to give each family an identification mark which will allow them to be recognised easily if for any reason they have to be segregated, or if a lamb loses sight of its mother and has to be reunited with her.

This is also a good opportunity to inspect each ewe's feet and trim if necessary, and also to give a dose of an anthelmintic to make sure she goes out worm-free. This dosing should continue at monthly intervals until the lambs are weaned.

During the lactation, and especially after weaning, keep a careful watch on the ewes for any sign of mastitis in overstocked udders, but do not be tempted to strip milk from ewes with healthy udders after weaning. This will only encourage the udder to produce more milk and will hold up the natural process of reabsorption.

If the ewe has successfully reared two or more lambs to weaning, she will very probably have lost condition and have a body condition score of around 2.5. Provided that she is healthy and that there is sufficient time to bring her up to a score of 3.5 before mating again, this is no cause for concern.

Before making up the breeding flock for the following year, a careful inspection of all the existing flock members is required to make sure that no unproductive passengers are being carried. Start with a look at the incisor teeth. Are the teeth present compatible with the age of the sheep? Do any sheep over four years of age have any incisors missing? If they have, and in all other respects they are sound and healthy, will facilities be available for them to have sufficient supplementary food in order to maintain their body condition, especially in the later stages of pregnancy? Feel the outside of the jaws along the site of the molars (back teeth) for any hard lumpy tissue which will indicate infected and possibly decaying teeth which will require veterinary attention. Next handle the udder, feeling for any lumps of hard tissue which will indicate chronic mastitis. A few weeks after weaning the udder should have returned to a soft, pliable consistency.

Any ewes with one or more of these defects or any adults that have not bred lambs when given the opportunity to do so for two consecutive years, should be given a separate identifying mark for culling. This also applies to any suffering from recurring chronic illnesses and those with chronic foot problems. Any ewes which have required a caesarian section operation to deliver their lambs will have only a fifty per cent chance of breeding successfully in future years, and their retention as flock members should be carefully considered.

Following this inspection the sheep marked as culls can be marketed for meat when the price offered is favourable. Any surplus ewes suitable for breeding can be advertised for sale or taken to sheep auction sales which take place from mid-August until November.

Final decisions on the size and composition of the flock can then be made a few weeks before mating is due to begin. Any additions or replacements will have been made well in advance.

SELECTION AND CARE OF THE RAM

The breed of ram selected to be the sire of the flock is one of personal preference, but certain basic requirements in the conformation are common to all. Look for:
1. Good feet and straight legs with an easy walking action
2. Correct mouth with neither undershot nor overshot jaw
3. Bold eye and alert expression
4. Large, even-sized testicles, firm to the touch
5. Long, straight back and good spring of rib
   Pay particular attention to the choice of ram, looking for one at the breed standard for size and weight. Always buy the best you can afford. If the ewes of the flock have a particular fault which needs correcting, beware of using a ram which is particularly good on that point while weak on others. Remember that the ram is responsible for

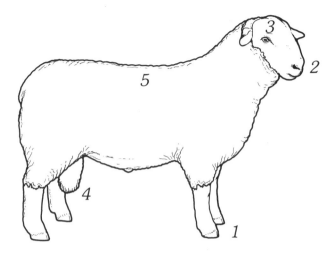

POINTS TO LOOK FOR IN A GOOD RAM

half of the genetic make-up of his progeny.

Try to arrange for the ram to be kept separate from the rest of the flock until he is required for mating, but if possible give him company by providing him with a wether, or a ewe that you know has been successfully mated. A ram kept by himself without the company of other sheep can become aggressive, although this aggression can be minimised by fitting a face mask which restricts forward vision. Avoid making a pet

Ram mask

of the ram by giving him titbits, as this too will encourage aggression.

Use a harness when the ram is with the ewes. The mark of a coloured crayon fitted to the harness will identify each ewe when mating takes place, and will let you know when to expect the lambs to be born. The harness should be fitted to the ram without the crayon a few days before mating is due to start to allow it to settle into the fleece, and if necessary it can be tightened to give a good firm fit when the crayon is added. Check periodically that the harness and crayon-holder are not chafing.

The crayon should be changed after seventeen days to another colour as a check that mating has been successful, as this is the interval when the ewe's heat cycle recurs. If more than one or two ewes are marked by the second colour, having already been marked by the first one, it may be that the fertility of the ram is questionable, but wait a few days longer to see whether more ewes return for a second service before considering whether it is necessary to use another ram. Another change of colour should be made after a further seventeen days, by which time all the flock should have been mated.

The ram should always be kept in good condition, fit but not overfat (condition score 3.5 to 4), and in good health. Pay particular attention to the feet, inspect regularly and trim as necessary. A lame ram will be reluctant to work and this can mean a reduced lamb crop.

It is difficult to say with accuracy how many ewes a ram can successfully serve in a season, but an optimum number is around forty-five. Mature rams from very active breeds like the Wiltshire Horn are capable of serving many more. Avoid overworking ram lambs.

If the size of the flock does not justify the cost of purchasing a ram, it may be possible to hire one just for the mating season, or to have an arrangement with another flock owner where the ewes can be taken. Bringing other sheep in or sending sheep away for a short time can bring problems, e.g. footrot and parasitic infection. Have a careful look before exposing the sheep to the possibility of infection from outside sources.

THE RAM HARNESS

37

CHAPTER 7   **Lambing**

Lambing time can be the most rewarding, demanding and sometimes the most stressful time of the year for both sheep and shepherd. It is the culmination of all the efforts to ensure that the health, nutrition and general wellbeing of the sheep result in a successful increase in the size of the flock.

By far the largest proportion of ewes lamb successfully unaided but a small number do need assistance. The way in which this assistance is given can vary a great deal, but the methods described in this book have been developed and proven over many years by us to give acceptable results.

It will not be possible to save every lamb that is born, as some may be still-born, but the golden rules are:

- Have your lambing kit to hand well before lambing commences, and also the telephone number of your vet.
- Practise as high a standard of hygiene as possible under the circumstances prevailing.
- Be patient. Do not rush to assist even in a normal birth, especially with ewes lambing for the first time. It is possible for the whole process to last for four hours and still result in a normal, natural birth.

The lambing kit: (1) lubricant; (2) baby feeding bottle and teats; (3) stock marker; (4) calcium borogluconate; (5) navel dressing; (6) pessaries; (7) torch (8) lambing ropes; (9) castrating pliers with rubber rings; (10) frozen colostrum; (11) stomach tube with syringe; (12) hypodermic syringe with needles; (13) vaginal support; (14) sheep halter; (15) lambing aid

THE LAMBING KIT

The lambing aids which you may require include the following suggestions:

● *A lubricant*   This is for use when it is necessary to assist a ewe by an internal examination and/or manipulation of the unborn lamb. This can be in the form of a mild antiseptic oil or cream, or soap and water. Soap flakes are particularly good.

● *A 1m length of thin cord*   By tying a loop at each end this can be used to secure an unborn lamb's feet or head and will make it easier to pull the lamb from the ewe should this become necessary. The cord should be stored in a weak antiseptic solution in an unbreakable container.

● *A lambing aid*   This is a Y-shaped plastic tube containing a retractable plastic covered cord. It can be used, with experience, to secure the unborn lamb's head and feet together inside the ewe and therefore enable the operator to withdraw the lamb from the ewe in a normal presentation position.

● *A navel dressing*   This dressing dries the lamb's navel very quickly and thus prevents harmful organisms gaining entry to the lamb, particularly the bacteria that cause joint-ill and liver abscess.

The dressing can be made from a tincture of iodine consisting of:
25g iodine
25g potassium iodine
25ml distilled water, freshly boiled and cooled. This mixture should then be made up to 1000ml with ninety per cent alcohol *or* with a very weak solution of copper sulphate.

Ideally, the dressing should be kept in a wide-necked container so that the navel cord and surrounding area can be immersed by holding the lamb vertically by the front legs and pressing the container gently against the lamb.

An alternative dressing is an antibiotic or antiseptic spray, which can be obtained from your vet.

● *A supply of colostrum*   Colostrum is the substance present in the udder when the lamb is born and, if the ewe has been vaccinated appropriately, it will contain the necessary antibodies to protect the lamb in the early stages of its life. The supply can be obtained from a ewe whose lambs have died or from one side of the udder of a ewe that has produced a single lamb. If neither of these sources is available then colostrum from a cow can be substituted although the lamb will not get the same protection from disease as from colostrum from a vaccinated ewe. It can be frozen in small quantities of about 200ml but it must be properly thawed and raised to body temperature before use.

● *A stomach tube with syringe*   This is necessary when giving colostrum to weakly lambs.

● *Baby feeding bottles and teats*
Sometimes the ewe's supply of milk in the first few days is insufficient for her lambs. If so, after the initial feed of colostrum, the lambs will need supplementary feeding with milk from other ewes or with a good-quality ewe milk replacer.

● *Rubber rings*   These are used together with the appropriate pliers for tailing and castrating.

● *Pessaries*   These are either antibiotic or antiseptic and are placed in the uterus of the ewe after her lamb has been delivered following internal manipulation. They can be obtained from your vet.

● *A hypodermic syringe and needles*
These are necessary for injecting the ewes and lambs.

● **A quantity of magnesium fortified calcium borogluconate** This is injected subcutaneously into ewes which collapse suddenly with hypomagnesaemia (magnesium deficiency) or hypocalcaemia (milk fever).

● **A wire or plastic support for vaginal prolapse** A vaginal prolapse usually occurs during the few weeks before lambing. Normally it first appears as a pinkish rose-shaped object, and may not be showing all the time. If the condition deteriorates and the prolapse gets bigger, then the ewe may have difficulty in urinating. This causes strain, and in the extreme condition the wall of the vagina may break and consequently the intestines escape through the hole. When it reaches this stage there is no cure and the vet should be called to destroy the ewe. The wire support, which has loops to secure it to the fleece of the ewe, should be inserted into the vagina when it is apparent that the prolapse is not going to go back on its own. It can be left in position until lambing is well advanced and the lamb is about to be born.

● **Stock markers** It is advisable to have several different colours available to mark lambs or ewes.

● **A halter** Occasionally a ewe will reject one lamb of her twins or triplets and a short spell tied with a halter will usually overcome this problem. The halter rope should be tied to a corner of the pen, high enough and short enough to prevent the ewe from turning round, but still allowing her to feed and drink easily. It is also a good idea to put a length of rubber tubing on the part of the halter which passes under the jaw. This will prevent the halter from becoming tight and therefore causing chafing.

● **A torch**

LAMBING: SIGNS AND OBSERVATIONS

The first indication that a ewe is about to commence lambing is that she behaves abnormally. Spend time watching normal behaviour in ewes so that any abnormal behaviour can be recognised.

The ewe will select a spot for her lambs to be born, and if she is driven away from it she will return at the earliest opportunity. She will become agitated, often pawing the ground and bleating softly. She will also lie down and get up again at frequent intervals. It is advisable to allow the ewe to remain at the site she has chosen, provided that it is sheltered, safe and dry. She is more likely to get on with the job of producing her lambs

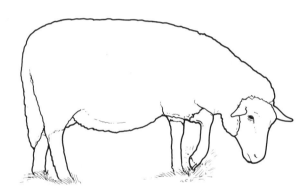

FINDING A SPOT TO LAMB

THE EWE STRAINS

Early signs

Selecting a spot

The ewe begins to strain

than if she is moved to another site or put in a small pen. Some ewes do not like an audience at close quarters when lambing and may give up attempting to lamb if anyone is near.

Labour

The ewe begins to strain very obviously as the birth commences and she lies down, often with her legs stretched out to one side. She often raises her head, closes her eyes and curls her top lip. The 'waterbag' is a transparent membrane containing the birth fluid. It may burst inside the ewe, but usually appears outside and bursts within a few minutes.

CURLING HER LIP

THE WATERBAG APPEARS

A normal presentation

With continued straining the lamb's nose and both front feet will appear. Often they will be enclosed in a transparent membrane. Slowly the body of the lamb will slide from the ewe. If the transparent membrane is still intact, prompt action is necessary to remove it; otherwise the lamb will be unable to breathe and will suffocate.

The nose and both front feet appear

A NORMAL PRESENTATION

The lamb slides from the ewe

The lamb will kick and cough and then the ewe will stand and turn to lick it. This has the combined effect of drying the lamb, establishing a bond between lamb and ewe and stimulating the lamb's circulation. This is a time of rapid heat loss in the lamb, particularly in inclement weather, and careful watch should be kept to see that it does not suffer from hypothermia due to exposure. If the ewe is reluctant to start licking, place the lamb at her head and rub some of the birth fluid over her nostrils.

After the birth of the lamb, the ewe can be moved to a sheltered position by holding the lamb by front or back legs and drawing it slowly in the direction that you wish the ewe to travel. It is very important that the lamb stays at ground level where the ewe can see it and can continue licking and smelling it as

The ewe licks the lamb

she moves. If she loses sight of the lamb she will return to the spot she chose for the birth.

Before leaving the ewe with her lamb, gently draw the thumb and forefinger down her teats to break the seal at the end so that the lamb can easily obtain the colostrum. The afterbirth should come away from the ewe after the birth(s) have been completed. Some ewes will attempt to eat the afterbirth, but it should be cleared away as it can distract the ewe from caring for her lamb. It will also attract predators — dogs, foxes and carrion crows. If the afterbirth becomes covered with straw and is allowed to decompose for some time, the smell of decomposition can be most unpleasant. It should be properly disposed of by burying.

MOVING THE EWE AND NEWBORN LAMBS

CHECKING THE UDDER

THE AFTERBIRTH COMING AWAY

Whether to assist the birth

Patience is a virtue! Do not rush to assist the ewe even in a normal birth. If, after an hour or so of labour commencing, nothing can be seen of the lamb and no progress is being made, then quietly catch the ewe. Insert one or two fingers into the vagina and try to identify any part of the lamb which you can touch. If an assistant is available to hold the ewe, this can be done while the ewe is still standing. Otherwise, it will be necessary to immobilise her by laying her down on her right side. This can be done by standing or kneeling at her left side and pulling her head round towards you while simultaneously reaching under her and grasping her right hind leg, then pulling it through in front of her udder towards you.

This is a move that can be practised at any time of the year, prior to the lambing season. Use a small sheep and progress to a larger one as you become proficient. The ewe will collapse onto her right side and your left hand and left leg can be used to keep her immobilised while the right hand makes a preliminary examination.

Remember that a normal birth can take up to four hours and if your examination shows that no part of the lamb can be felt and entry to the uterus is dry and restricted, then release the ewe and wait a while before making another examination. If after three hours no further progress is being made, **do be patient** and **do not force** an entry. Seek professional help.

Ringwomb

This is a condition where the cervix, or entrance to the uterus, fails to dilate. It can be recognised when a preliminary examination reveals, as the name implies, a tight inelastic ring at the entrance to the uterus. Do not force an entry but contact your vet.

IMMOBILISING THE PREGNANT EWE

KEEPING THE EWE IMMOBILISED

Dead lambs

You may find that you are having to assist in the birth of a dead lamb. This can be a difficult and unpleasant task, especially if the lamb is beginning to decompose. The dead lamb will feel dry to the touch and the natural lubrication of the birth will be absent. Use copious amounts of lubrication, soap and water for preference. If you feel you are unable to cope with this, consult your vet. He will also advise you on the use of long-acting antibiotics and pessaries.

Malpresentations

Unfortunately a number of malpresentations can occur, and if the examination shows that the lamb is being presented in a position other than normal, then one of the procedures on the following pages is called for.

One front leg back   This position means that only one front leg is forward and that the other front leg is lying alongside the lamb's body inside the ewe. To deliver the lamb, the visible hoof should be gripped firmly and the leg pulled out of the ewe in line with her backbone. Run one or two fingers around the top of the lamb's head to ease it from the vagina. When this has been done, pull the leg and the head towards the ewe's hocks, maintaining an even pressure until the lamb has been delivered.

ONE FRONT LEG BACK

THE VISIBLE HOOF SHOULD BE GRIPPED FIRMLY

PULL THE LEG AND HEAD TOWARDS THE EWE'S HOCKS

BOTH FRONT LEGS BACK

EASE LAMB'S HEAD OUTSIDE THE EWE

Both front legs back   In this case ease the lamb's head right outside the ewe and slide the right hand down the left side of the lamb's neck, feeling for its left shoulder and left front leg. Crook a finger behind the leg and pull it up in line with the lamb's neck. Deliver the lamb by pulling the foot and head downwards towards the ewe's hocks. It is not necessary to re-insert your hand into the ewe to bring up the other front leg.

If the ewe has been unattended and in labour for some time, she may have succeeded in getting the lamb's head out of the vagina. When this happens its head and tongue may appear very swollen but the swelling will quickly subside once the lamb has been delivered.

LOCATE A FRONT LEG AND PULL TO THE OUTSIDE. DELIVER AS WITH ONE FRONT LEG BACK

HIND LEGS FIRST

Hind legs first   If, on examination, only two feet can be felt and no head, and the points of the hooves are on the underside, then feel up the legs for the lamb's hocks. If they are present grip them firmly and quickly deliver the lamb by pulling it out in line with the ewe's backbone. It might be necessary to put a cord around the feet of the lamb to enable you to get a better grip on them, but it is important that the lamb is delivered quickly. If not, there is a possibility that it will attempt to breathe when the umbilical cord breaks while the head is still inside the ewe.

It will then take the birth fluid into its lungs and may suffocate. If you suspect that this has happened, hold the lamb by its hind legs and swing it violently backwards and forwards in an attempt to expel the fluid and mucus from its lungs. Clear away any mucus from the lamb's mouth and check whether its heart is beating by gently pressing on its rib cage immediately behind its shoulder. If the heartbeat is present but the lamb is not breathing, mouth-to-mouth resuscitation may help. If the need arises, don't be put off by the appearance of the lamb — there is virtually no taste to the birth fluid!

PULLING THE LAMB OUT HIND LEGS FIRST

TAIL FIRST OR BREECH PRESENTATION

Tail first or breech presentation   An indication that the ewe is attempting a breech birth will be the appearance of the lamb's tail. Gently push the hindquarters of the lamb back inside the ewe and reach for its hind legs. Pull both hind legs up and outside the ewe, and then quickly pull the lamb out in line with the ewe's backbone. If necessary, swing it and proceed as with hind legs presentation to expel fluid and mucus from the lungs.

DEALING WITH A BREECH PRESENTATION

48

Head back, both front feet forward This is the most difficult presentation to deal with. Firstly, identify positively that the feet are front feet by noting that the points of the hooves are uppermost and feeling for a knee. Do not under any circumstances pull on these legs hoping that a head will follow.

If the head is turned back, pulling the legs will force the head even further back. Instead, put the lamb's feet in the palm of your right hand and push them back inside the ewe, and as soon as it is possible to do so, push them down so that the legs are under the body of the lamb. Locate the curved body of the lamb and pull the head forward by holding the lower jaw between the thumb and forefinger, or by putting the thumb and forefinger in the eye sockets. Bring the head right outside the ewe and slide the hand down the lamb's neck, bringing up one front leg, and delivering the lamb by pulling its head and leg towards the ewe's hocks.

This is an occasion when the lambing aid comes into its own. It can be used to secure the lamb's head after locating it inside the ewe, and also for pulling the head through the pelvis to the outside.

Twin lambs presented together This can be a puzzling situation for beginners. If two heads and some or all of the front legs are presented together, it will be necessary to push one head and all the legs back inside the ewe. Bring one head to the outside before going back for one front leg, making sure that it belongs to the same lamb as the head! Deliver the lamb by pulling the head and feet downwards towards the ewe's hocks.

In all other cases of mixed-up presentations, try to identify positively any part of a lamb that you can touch. If a head is presented, then work with this one. If at any time it is difficult to move the lamb inside the ewe, it sometimes helps to roll the ewe over onto her right side and then back again to the left.

HEAD BACK AND BOTH FRONT FEET FORWARD

TWO METHODS OF BRINGING THE HEAD FORWARD

Guarding against infection

In all cases of assisted birth an antiseptic or antibiotic pessary should be inserted into the ewe when all the lambs have been delivered. Additionally, an intra-muscular injection of a long-acting antibiotic should be given to prevent infection. This is obtainable from your vet.

CARE OF THE LAMB

The first few days of a lamb's life are critical to its future development, and a careful watch should be kept on all newborn lambs so that any potential problems can be recognised early.

Following a successful birth the ewe normally begins to lick the lamb clean and dry. The lamb gets to its feet after a few minutes, makes it way to the ewe's udder and begins to suck vigorously. The ewe and lamb should be placed in a small pen (1.8m × 1.8m) on a clean bed of straw. This will prevent the lamb from straying, especially in the case of a multiple birth where the ewe might have to care for several lambs. The lamb's umbilical cord should be dressed (see the lambing kit list) to guard against infection.

If the lamb has been born outside in inclement weather, it may be suffering from hypothermia, or chilling, and finding it difficult or impossible to get to its feet. Help to stimulate circulation by rubbing the lamb briskly with a dry towel, and introduce it to the ewe's milk by opening its mouth and inserting a teat, gently squeezing it to give the lamb a taste. This task is made easier by casting the ewe onto her side, and then by leaning over her and placing an ankle over her neck, which will leave both hands free to handle lamb and teat. An elbow gently placed into the flank of the uppermost leg of the ewe will prevent a kick in the face! If the lamb is reluctant to suck, it can sometimes be stimulated by scratching its tail head.

If it is still cold, wet and lethargic but able to hold up its head unaided, it can be fed using a stomach tube. To do this, sit comfortably with the lamb on the lap. Introduce the tube minus the syringe into the lamb's mouth and gently slide it into the stomach, making sure that it is in the oesophagus and not the windpipe. If the lamb shows no sign of discomfort and chews the tube when it is in position, this indicates that it is in the stomach. In a large lamb all

DRESSING THE UMBILICAL CORD

but two inches of the stomach tube can be introduced. The syringe should be filled with colostrum taken from the ewe and placed on the end of the tube. Depress the plunger of the syringe slowly and repeat the process until the lamb has received between 100 and 200ml, repeating this four hours later if necessary. As soon as the lamb begins to suck unaided, place it at the ewe's teat. Ensure that the lamb gets an adequate supply of at least one litre of colostrum during the first twenty-four hours. If the ewe has no milk, use a baby's feeding bottle for giving a good quality ewe milk replacer made up in accordance with the manufacturer's instructions. Apply a strict hygiene routine to the stomach tube, syringe, bottle and teat used to feed lambs, and thoroughly clean and sterilise between each feed.

FEEDING WITH A STOMACH TUBE

If it becomes necessary to warm a lamb by an artificial heat source, several methods can be used. An infra-red lamp is useful, but take care to see that it is far enough away from the lamb (a minimum of 1.2m) to prevent scorching or burning. A drawback to this method is that when dry the lamb takes on a different smell and because the mother of a newborn lamb identifies it by smell alone she sometimes refuses to accept that it belongs to her. A way around this problem is to suspend the lamp above the lamb in a small fenced off corner of the pen where the ewe can still see and smell the lamb but where she cannot be burned.

A domestic hot air blower can be used to blow warm air onto the lamb lying in a box or confined space, as can the oven of a domestic cooker at a low temperature, but remember to leave the oven door open!

If no artificial heat source is available, a hay box can be used to revive a lamb suffering from exposure. Use soft hay to make a deep nest inside a container such as a large dustbin or forty-five gallon drum. Place the lamb in the nest and cover with clean hessian sacks or more hay before putting a lid on the container. An hour or so

FEEDING WITH A BOTTLE

51

later the lamb will usually be standing on the sacks or hay and calling for food.

Any lamb which is cold and lethargic when several hours old and already dry should not be subjected to artificial heat. Give a feed of colostrum from a baby's feeding bottle or through a stomach tube if the lamb is reluctant to suck.

Castrating

When it is apparent that all is well — the lambs are receiving an adequate supply of milk and the ewe is mothering properly — the ram lambs can be castrated. The simplest way for the beginner to do this operation is to use rubber rings and the appropriate pliers, called elastrators. By law this method is permissible only during the first week of a lamb's life, and by persons over seventeen years of age. First check for abnormalities. The lamb may be suffering form a scrotal hernia, when a part of the intestine is present in the scrotum with the testicles, and can be detected as a soft, spongy mass. Lambs with this condition will die if castrated. Sometimes either one or both of the testicles have not descended into the scrotum. If only one has descended, do not castrate. The tail should be left long while waiting to see whether the testicles come down normally.

2

3

STEPS IN CASTRATING

1

4

Kneel or sit with the lamb lying on its back between your legs and facing away from you. Place a rubber ring on the elastrator, stretch the ring open and draw the scrotum through it. Close the ring on the scrotum and then open it just sufficiently to allow a testicle to be eased through it into the scrotum, using the free hand (see illustrations on facing page). Close it again and repeat the procedure for the second testicle. Make sure that the rubber ring is well below the rudimentary teats and that both testicles are secured before removing the elastrator from the ring.

At sixteen to twenty weeks of age an uncastrated ram lamb becomes sexually active and can be a nuisance if left with the flock. Either segregate it or send it to market.

If it becomes necessary for any reason to remove the ring from the scrotum, insert a small piece of wood under the ring and cut through the ring onto it. This will avoid cutting the skin.

Docking

To dock the lamb place a ring on the elastrator, open the ring and pass the tail

DOCKING

through it. Before closing the ring on the tail, ensure that sufficient tail is left above the ring to cover the anus of ram lambs and the vulva of ewe lambs. (This is required by law.)

After these operations both ram and ewe lambs suffer some discomfort and will roll around on the ground for a short time. Because of this it is advisable to do the job either on a clean straw bed or on a clean grass field. The discomfort soon passes and the lambs are soon behaving normally again.

The correct tail length

FOSTERING LAMBS

Occasionally it may be necessary to foster an orphan lamb onto a ewe that has lost her lamb, or to foster a lamb from a multiple birth onto a ewe that has produced only a single lamb. There are several ways of doing this.

Skinning

In the case of a dead lamb remove the skin, trying if possible to leave the holes for the legs intact. Place the skin with the tail still attached onto the lamb to be fostered, putting the legs through the holes left in the skin so that it covers all the body except the head and ears. If necessary, use string to join the skin to make a good fit. As a safeguard, put a halter on the ewe and tie her to the corner of a small pen. Tie the rope at a height that will allow the ewe to lie down in comfort but will not permit her to turn round or step over the rope. She will usually accept the lamb straight away, but it may be necessary for the first few days to shorten the halter rope every few hours and stand by while the lamb is encouraged to suck.

Using birth fluids from a newborn lamb

To foster a lamb onto a ewe that is producing a single lamb takes rather more expertise. Firstly, be present at the birth. Immediately afterwards, before the ewe stands up or sees her own lamb, press a hand onto the abdomen just in front of the udder to see whether there is another lamb in the uterus. If there is, it can be felt as a firm, solid mass and as an additional check put one finger into the vagina to feel for any lamb about to be born. If no other lamb is present, keep the ewe in a lying position and if necessary wrap her head in a cloth so that she is blindfolded. She will then lie quietly.

Collect the lamb to be fostered, and using string, tie a front leg to the diagonally opposite back leg so that it is unable to stand. The foster lamb will usually be dry, but if it too is newborn and still wet, rub it briskly with a dry towel to dry it as much as possible. Use all the birth fluids available from the newly lambed ewe and her lamb, if necessary squeezing fluid from the legs and nose of the lamb, to saturate the foster lamb. Place the foster lamb at the ewe's head and remove the blindfold. Take away the ewe's own newborn lamb for about twenty minutes while the ewe licks the foster lamb, before returning it to its mother. The attempts of the foster lamb to stand while its legs are tied simulate the efforts of the newborn lamb trying to get to its feet. Wait another twenty minutes before untying the legs. During this time the newborn lamb, if it is healthy and normal, will have got to its feet and taken its first feed of colostrum from the ewe. Keep the ewe and lambs in a small pen for a few days to check that all is well and that both lambs are thriving.

Lamb adopters

These are boxes or crates which constrain a ewe in such a way that she cannot see or smell the lamb which is being fostered. The lamb has access to the ewe at all times and also has a safe area where it cannot be trampled on or kicked and butted by the ewe. The apparatus can be purchased or made up at home using plywood. It consists of a stall approximately 1 metre wide and 1.2 metres long, with a slot at the front just wide enough to take the ewe's neck comfortably, but narrow enough to prevent her from withdrawing her head when placed in it. A rail on either side of the ewe prevents sideways movement and provides the lamb with a safe, draughtproof area.

The theory is that after a few days when the milk has passed through the lamb, it will take on the smell of the ewe and that she will eventually recognise and accept it as her own. This method is more successful with ewes that have only recently lambed. After release keep both ewe and lamb in a small pen until it is seen that all is well.

Ewes which are normally placid creatures can behave violently towards lambs which are not their own. Sometimes a ewe which has given birth to twins will reject one for no apparent reason. If both lambs are strong and active, tie the ewe to a corner of the pen using a halter, and after a few days she will usually accept both lambs. Try taking both lambs away for about half an hour. When they are returned she will usually be so pleased to see them that she allows both of them to suckle her.

After tailing, castrating and any fostering have been successfully completed, several ewes with their lambs can be grouped together in a small enclosure. Use a stock marker aerosol to identify lambs with their mothers by marking them on both sides of the body with a similar number or symbol.

From now on, given an adequate supply of milk and grass, the lambs should thrive. Dose with a broad-spectrum anthelmintic at six weeks of age and again at monthly intervals to keep parasitic worms under control. If supplementary feeding is needed for the lambs, it can be started when they are around ten days of age. It can be in the form of pellets specially formulated for lambs or as a coarse mix containing ingredients which the lambs find particularly attractive — a kind of lamb muesli! Both will be available from your feed merchant. From an early age the lambs will also eat hay which, as is the case for all classes of sheep, must be of good quality.

The feed for the lambs should be made available to them via a lamb creep through which the ewes cannot pass. It should be offered to them ad lib until they are eating around 300g a day. After that they should have a single feed each morning, the amount being dictated by the quality and the quantity of the grazing and the condition of the lambs. The feed, particularly when

A thriving lamb

offered ad lib, should be protected from the weather and the lambs prevented from standing in the troughs when feeding.

Clean water should always be available to both ewes and lambs.

All the sheep are at risk of fly strike between April and November when they should be checked at least twice a day (see blowfly strike in Chapter 8). Short-term protection can be provided by dipping in an insecticidal dip.

If lambs are intended for the meat market, periodic weighing is required. With small numbers this can be done using bathroom scales or a sling or container and spring balance. The optimum weight for marketing is between 32 and 40kg liveweight which will give a carcase of approximately 50 per cent of that weight, depending on the degree of finish or fatness.

HAND REARING ORPHAN OR SURPLUS LAMBS

Day 1   It is absolutely essential that newborn lambs are given colostrum as their first feed. Colostrum is a sticky, yellow substance present in the udder when the lamb is born. It is a rich source of energy; it is a laxative; and provided the ewe has been given the appropriate vaccines, it will contain antibodies which give the lamb protection from specific diseases through the early stages of its life. Each lamb requires a minimum of one litre of colostrum during its first twenty-four hours of life, but as the ewe produces sufficient for only two lambs, it follows that any surplus lambs from multiple births, triplets or quadruplets, will need a supply collected from another source.

Probably the best source is from ewes which produce single lambs. One side of the udder can be hand-milked into a container and the colostrum deep-frozen in 200ml packs. Other supplies can come from goats or dairy cows, which will require to be vaccinated beforehand if the lamb is to be protected by antibodies in the colostrum. The deep frozen packs need careful thawing before feeding and should be heated to the body temperature of a normal, healthy lamb which is 39 – 40 degrees C (102 – 104 degrees F). There are also commercially produced colostrum powders available for use if all else fails.

Bottle feeding

Days 2 to 9   Introduce a good quality ewe milk replacer of preferably 30 per cent fat content, mixed with water in accordance with the manufacturer's instructions. Feed warm in four feeds to a total of one litre per lamb per day, using a bottle with teat attached.

It is possible to rear lambs on ewe milk substitute fed cold. It should be available ad lib and should be offered via teats fixed to the side of the pen, connected by plastic tubing to a supply of milk outside the pen. It is very important that the teats when fixed in position are above the level of the milk substitute in the container. The lambs will need to be taught how to use this feeding method by holding them onto the teat for the first few feeds. An infra-red lamp suspended not less than 1 to 2 metres above the pen will be appreciated by them.

Days 10 to 19   Introduce hay and concentrate feeding. As always, the hay should be of top quality and in order to tempt the lambs to eat concentrates, offer a small amount of flaked maize initially. Small pellets of compound feed purchased from your feed merchant can be introduced gradually. Restrict warm milk to two feeds per day.

Days 20 to 42   Continue with hay and concentrate feeding. Provide clean water. The water container and any feed troughs used should be scrupulously clean and set up in such a way that the lambs cannot foul them with their feet or with urine or dung.

Day 43 onwards   The lambs can be weaned and turned out onto clean grass. Feed concentrates twice daily at the rate of 250g per lamb per feed. They will be ready for marketing when they reach 32 – 40kg.

CHAPTER 8   **Some Diseases Affecting Sheep**

Lameness

Lameness can be caused in a number of different ways. Watch the sheep to see how it moves and to establish which leg is lame. Catch the sheep, immobilise it and examine each foot in turn. Look for a reddened, moist area between the digits and also for abscesses around the top of the hoof. If these symptoms are not found, feel the leg above the hoof for any swelling or 'hot spot' which might indicate a sprain or hairline fracture. Check that there are no dislocated joints.

Footrot and Scald

Footrot is a very painful and debilitating disease if allowed to go untreated, and is directly responsible for causing huge financial losses to the sheep industry every year. While it is not a killer disease in itself, it can severely affect the performance of all classes of sheep from young lambs and ewes to rams, the latter especially at mating time.

It is caused by the joint action of several types of organisms, one of which is quite common in some areas of the United Kingdom. It causes scald — an inflammation of the skin between the two digits of the foot — and it can occur in all classes of sheep at any time of the year, but is more commonly seen when the sheep, particularly lambs, are grazing long, lush grass in the spring and also shortly after ewes have been housed on straw bedding in the winter. The particular organism which causes scald can only live for three weeks away from the sheep, and careful use of a footbath can eradicate it

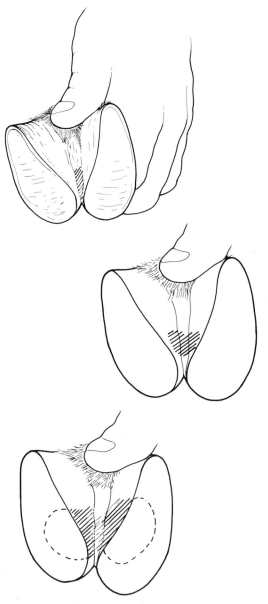

HATCHED AREAS SHOW THE SPREAD OF FOOTROT

58

from the flock, especially if clean land is available for grazing after footbathing.

True footrot occurs when a second organism invades the inflamed tissue between the digits. This causes an under-running of the sole of the foot, and eventually the horny outer wall, reaching right up to the toe. If this under-run tissue is lifted or pared off, the exposed area resembles the gills of a mushroom, except that it is usually dark grey in colour, and has a characteristic, pungent, unpleasant smell. Treatment consists of paring away all the under-run horn from the sole and outer surface of the digit. The bulk of the material can be cut away with straight-bladed secateurs, but a very sharp knife should be used for the more precise trimming necessary to expose the fullest extent of the infection. The dividing line between infected and healthy tissue is a very fine one, and every care should be taken to avoid the bleeding which will occur if the healthy tissue is cut, especially in the region of the toe.

Here it should be stated that if footrot is treated as soon as a sheep is seen to be walking lame, a lot of pain for the sheep and a lot of work for the shepherd can be avoided. An antibiotic spray can be used to treat the infection and the sheep should be stood on a hard surface while the foot dries.

When footrot or scald is present, it is much easier to control if all the sheep can be stood in a shallow bath of ten per cent zinc sulphate solution for up to an hour. This long period enables the solution to penetrate pockets where the causal organisms might have been missed when paring. The feet should then be allowed to dry before returning to pasture.

Formalin is also used for treatment and control, but it has a number of disadvantages. The fumes given off are toxic, and it is ineffective when the footbath becomes contaminated with dung or soil. It makes the hoof very hard and is very painful if used on feet where footrot has already been exposed. The feet should be soaked for a much shorter time with formalin solution, about two minutes usually being sufficient.

Both zinc sulphate and formalin are difficult to dispose of and both are poisonous to fish if allowed into streams, ponds or ditches. Zinc sulphate may be used and re-used as it does not deteriorate in contact with vegetable matter.

When treating footrot it is important that all foot parings are collected and burnt and that all tools are disinfected by immersing them in ten per cent zinc sulphate solution between each sheep.

In severe cases of footrot, and particularly with valuable rams, consult a veterinary surgeon about the possibility of an antibiotic injection as treatment after paring. Prevention can be assisted by using a footrot vaccine. Two vaccinations are required initially, followed by two booster injections each year. The vaccine is expensive to buy, and any partly used bottles must be discarded which adds to the expense for small flock owners.

Foot Abscesses

These can cause very painful lameness, occurring where the horn of the hoof joins the skin of the coronet. Consult your vet on treatment.

Granulomas or Proudflesh

When the feet are allowed to become badly overgrown, the horn may aggravate the soft tissue between the digits, or the under-run sole may rub against the sensitive layer below it and cause excessive growth with associated bleeding, especially from the area of the toe. This is a very painful condition and in the spring and summer may attract the blowfly to lay its eggs in the foot. (This is the case with ordinary footrot.) Granulomas need to be surgically removed by a vet and even so they take a long time to heal.

METABOLIC DISEASES

Pregnancy Toxaemia

This is a disease which occurs during the last month of pregnancy, most commonly in ewes that are carrying two or three lambs. At this time the unborn lambs are making heavy nutritional demands on the ewe. Pregnancy toxaemia can affect both thin and fat ewes and is usually caused by an insufficient uptake of nutritious food. Concentrates should be fed during the last six weeks of pregnancy, starting with 100g per day, increasing weekly to reach 750g per day by lambing. The protein content of the concentrates should be seventeen per cent, and it is helpful if the merchant supplying the feed gives you a list of the ingredients so that the stated analysis can be checked. Ewes affected by pregnancy toxaemia get separated from the rest of the flock, refuse food and become blind. After a few days they are unable to stand and they assume unnatural positional behaviour. Front legs may be stretched out in front and the head pressed down onto the ground. Symptoms include twitching and trembling of the lips and ears, and grinding teeth. If the ewe is left untreated, death follows fairly quickly. The advice of the vet should be quickly sought if pregnancy toxaemia is suspected, and the ewe moved to warm, comfortable conditions under cover.

Prevention of this disease is easier if sheep are housed, as a more careful check can be kept on their condition. The target condition score at lambing is 3.5, and thin ewes and shy feeders which are often bullied at the trough should be separated and given extra concentrates. Ewes which are lame or have teeth problems are also at risk, as are ewes kept outside in periods of very bad weather. See that all the ewes have sufficient trough space. It is helpful to be able to put all the feed in the troughs before allowing the sheep access to it, perhaps in a separate enclosure or an adjoining field. This will allow fair shares for all.

The bulk part of the diet should be of good quality hay, and the concentrate fed first. A stomach full of poor quality food is of no use to a ewe carrying three lambs. Four-fifths of the eventual birthweight is gained during the last two months of pregnancy.

Overfat ewes too are at risk, simply because internal layers of fat restrict the space available for any increase in nutritious food required to feed the unborn lambs.

Death is caused by a breakdown of the metabolism and a lowering of blood sugar levels. One school of thought suggests that if the ewe is shorn the metabolism is stirred into action again by the demands of the ewe as she attempts to convert energy into warmth! It is better to prevent the disease by careful management of the flock, particularly in the later stages of pregnancy, as the recovery rate is very low.

Hypocalcaemia

This disease can occur during late pregnancy or early lactation, and it is caused by a low level of calcium in the blood. Listlessness and depression precede loss of consciousness, but a subcutaneous injection of 100ml of twenty per cent calcium borogluconate will effect a rapid return to normal. Divide the dose and inject at several different sites.

Hypomagnesaemia

The onset of this disease is extremely rapid and usually occurs when newly lambed ewes are turned out onto very lush grass. It is caused by low levels of magnesium in the blood, and is characterised by a stiffness when walking, and twitching and spasms, followed by convulsions and death. Prompt treatment, consisting of a subcutaneous injection of 100ml of magnesium fortified twenty per cent calcium borogluconate divided and injected at several different sites as with hypocalcaemia, is necessary. The main problem in treating this disease is the speed with which it develops, and unless the person in charge is present and very observant, quite often the only indication is a dead sheep. Prevention can be assisted by offering a mineral mixture containing calcined magnesite, but as this is an unpalatable ingredient it needs to be disguised to make it acceptable by adding some molasses or treacle. It is possible to buy compound feeds containing calcined magnesite, and these should be introduced into the rations gradually just before lambing, and continued for a few weeks afterwards while animals are at risk.

Heavy dressings of nitrogenous fertilisers encourage a rapid growth of young grass in the spring which also reduces the uptake of both calcium and magnesium, a point to be borne in mind when considering how best to manage the available grassland.

Swayback

Swayback is another disease caused by a deficiency, this time of available copper in the bloodstream of ewes in the later stages of pregnancy. It occurs in localised areas, and local sheep farmers and the vet will be able to tell you whether there is a history of copper deficiency in your area. Swayback can be recognised as a type of paralysis of the hindquarters in newborn lambs, making it impossible for them to stand. Otherwise they are bright, but they never recover and should be taken for veterinary examination so that an accurate diagnosis can be made. This disease can be prevented by an injection of a suitable preparation of copper halfway through the pregnancy and obtainable from your vet.

At this point, however, it must be stressed that the sheep is more prone to copper poisoning than any other farm animal. Some breeds are more susceptible than others, and housed sheep more so than

sheep on hill grazings. Small amounts of copper taken in over a long period can accumulate to the extent that even a correct dose of copper in a readily available form can prove toxic if the animal was not deficient in it at the time of dosing. Similarly, mineral licks high in copper and concentrate feeds which may contain more than fifteen parts per million should not be used. Concentrate feeds formulated for other forms of farm livestock are particularly dangerous as they contain a much higher level of copper than sheep require.

As stated earlier, discuss with your vet whether there is a copper deficiency in the area and be advised by him as to the correct compound to use, and also the correct dose for the various classes of sheep. Ensure that no other form of copper is available if the ewes are to be given injections of copper, and do not give anthelmintics for worm or fluke control at the same time.

VARIOUS OTHER DISEASES

Gid

This disease of the central nervous system is caused when the animal eats grass contaminated by the larval stage of a tapeworm found in dogs. Over a period of time the larvae migrate through the bloodsteam and begin developing when they reach the nervous tissue, including the brain. Once there they grow into a cyst filled with fluid. This contains tapeworm heads which develop into adult tapeworms if, after death, the head of the sheep is eaten by a dog.

The symptoms which may indicate gid can vary, and are easily confused with other nervous diseases. Sometimes the affected sheep will lie with the head turned to one side, or hold it abnormally high. Others will walk in circles, always turning the same way, and all will eventually go down, become comatose and die. The time taken for this to happen varies considerably, and if the veterinary surgeon is called in early and diagnoses gid, the chances are that he will be able to operate successfully and remove the cyst.

It is possible to break the life cycle of the parasite by ensuring that any sheep offal fed to dogs is boiled for at least one hour. All dogs owned should be dosed with a medication, obtainable from your vet, which is capable of killing all of these tapeworms, including the heads. This should be done every two months, but of course it will not prevent other dogs and foxes coming onto the property and contaminating the grassland. Another safeguard is to bury all sheep carcases or to make sure that they are disposed of in such a way that dogs will not eat the uncooked remains.

Listeriosis

The listeria bacteria lives in soil contaminated by the faeces of animals. If a large amount of soil is collected with the grass in the process of silage making, and especially if poor fermentation due to incomplete exclusion of air takes place, the bacteria multiply. If this silage is fed to sheep, they are at risk of infection.

The resulting illness can affect sheep in several ways. Encephalitis is caused by numerous abscesses on the brain. In this form the sheep's head is inclined to one side with the ear and eyelid on that side drooping, and with saliva dribbling from the mouth. It may fall and be unable to rise, become comatose and die, all within two days. Lambs can be affected through contamination with the soil via the wet navel cord, by nibbling at silage or by coming into contact with aborted material, the abortion having been caused by listeria, and diarrhoea being the principal symptom.

It follows that if silage is not fed, the risk of listeriosis is greatly reduced.

Clostridial Diseases

There are seven related clostridial diseases which can affect sheep of all ages, most of which are characterised by sudden or rapid death. Fortunately all can be controlled by using a single vaccine which can be given to lambs as a subcutaneous injection at twelve weeks of age, followed by a second dose six weeks later. A booster dose is given around twenty-one days before lambing, and when the lambs are born they are protected by antibodies in the colostrum for ten weeks. In the case of non-breeding sheep they can be injected at a similar time of the year. It is a very cheap vaccine and an essential part of sheep husbandry, the loss of one unvaccinated ewe being equivalent in cost to 350 doses of vaccine.

Pasteurellosis

The bacteria which cause pneumonia in sheep come in many different types. Deaths from pneumonia can occur in all classes of sheep and at any time of the year. Often the first sign of the disease is a dead sheep, or a short period of laboured breathing and frothing at the mouth followed by death. Other forms of the disease can be treated with long-acting antibiotics if detected early.

Pasteurellosis is a very complex and variable condition. Periods of stress, which can include bad weather, can predispose a flock to infection, and if the level of stress is great enough outbreaks can occur even though the sheep have been vaccinated. A post-mortem examination of sheep which die from pneumonia will reveal plum-coloured lungs surrounded by bloodstained fluid, and in many cases death will have occurred so rapidly that it is impossible to observe any symptoms.

It is possible to vaccinate against certain types of pasteurella, and to be effective the vaccine must be appropriate to the particular requirements of the flock as determined by laboratory tests. It is possible to buy a combined clostridial and pasteurella vaccine, but little, if any, immunity to pasteurella is passed to the lambs in the antibodies contained in the colostrum, compared to the cover given against the colostridial diseases. If early protection is required for the lambs, they must be given pasteurella-only vaccine when two weeks old if they are born to ewes which have previously been vaccinated.

Lambs born to ewes which have not previously been vaccinated can be given pasteurella-only vaccine at birth, followed two weeks later by a second injection. Ewe lambs retained for breeding, will require one annual booster injection.

Mastitis

Mastitis can be present in the ewe flock in both chronic and acute forms. The chronic form is usually discovered when the ewes are being inspected prior to selection for the breeding flock. It is identifiable as a fibrous lump or series of lumps in the udder; and if the ewe is used for breeding, the affected side of the udder will not produce any milk. It is not known when infection occurs and there is no known cure. Affected ewes should be culled.

Mastitis

The acute form commonly occurs early in lactation. Affected ewes are clearly ill and appear to be walking stiffly, while the udder is swollen and grows increasingly cold to the touch. The udder rapidly becomes inflamed and red or purplish blue, and eventually black. If the teat is drawn, a bloodstained fluid comes away. The inflammation may extend along the belly of the ewe if she survives, and eventually all the dead, coloured tissue will fall off, leaving a raw, bleeding surface. Healing takes a long time and the exposed tissue is vulnerable to flystrike. Consult the vet as to the best course of action. There is no effective treatment for this condition, and if the ewe survives she should be culled.

Tickborne Diseases

Sheep ticks have a limited distribution in the British Isles, being principally confined to areas of rough vegetation on acid soils. Sheep in these areas acquire an immunity to some tickborne diseases, but problems can occur when unacclimatised sheep are brought in. Rams brought in from 'clean' areas can become infertile, and tick bite pyaemia, which is associated with painfully enlarged joints, can cause abscesses to develop in various parts of the body.

These infections emphasise the importance of effective tick control if sheep have to use tick-infested land. Drainage and pasture improvement will eliminate conditions favoured by ticks, and dipping in an appropriate dip in the early spring and again in the autumn will effectively reduce numbers. Preparations which are poured onto the sheep in a line from the head to the rump will give several weeks' protection.

Orf or Contagious Pustular Dermatitis

Orf is a very painful condition affecting sheep of all ages. It is caused by a virus which enters the skin through minor abrasions. It then multiplies rapidly, causing small spots which suppurate and join together to form grape-like clusters. It is confined to the non-woolly areas, and in young lambs it affects the lips, gums and nose. They in turn transfer the infection to the teats and udder of the ewes, resulting in the doubly distressing situation where hungry lambs with very painful mouths are denied access to the milk because of the soreness and pain of the ewe's teats. In a few weeks the condition clears up and healing occurs. Very little immunity is produced through the antibodies present in the ewe's colostrum at lambing.

Orf can also affect older weaned lambs in the autumn, and the genital areas of ewes and rams, making the rams very reluctant to work. The condition known as strawberry footrot, which causes raw red bleeding areas on the face and legs of older sheep, is believed to be related to the orf virus.

Being a virus, orf does not respond to antibiotics, and virtually nothing can be done to effect a cure. Secondary infection of the lesions can be guarded against by using antibiotic aerosols. It is possible to vaccinate against orf. This consists of applying the vaccine to an area of bare skin, preferably between a foreleg and the wall of the chest. Do not use disinfectant or spirit to clean the skin before applying the vaccine, which is made from the living virus, as it will be destroyed before producing the desired effect. The skin should be lightly scratched where the vaccine is applied, and later a scab will form.

Flocks which have already had an outbreak of orf should be vaccinated some two months before protection is required, but on no account should sheep be vaccinated in flocks where the disease has never occurred. This is because mild infection is being introduced by a modified form of the living virus, and the scabs as they fall from the sheep also contain the living virus and are liable to infect any 'clean' sheep brought in. An outbreak of the disease can be controlled to a certain extent, especially in young suckling lambs, by vaccinating non-affected members of the flock. Seek the advice of your vet if an outbreak occurs, and follow the manufacturer's instructions carefully when using the vaccine. Wear protective gloves at all times when dealing with this disease, as orf can also affect humans and cause very painful lesions on the hands, arms or face.

SOME PARASITIC PROBLEMS

Liver Fluke

The fluke is a flat, leaf-like parasite which is common in wet, marshy areas. It requires an intermediate host, a mud snail, for its development. For the sheep the problem starts when it eats grass which is carrying the encysted stage. On reaching the sheep's gut the immature fluke makes its way through the gut wall and searches out the liver. Once in the liver it feeds voraciously, and on becoming adult it anchors itself in the bile duct and takes blood from the wall. It lays eggs which pass out with the faeces. These then hatch and seek out a mud snail, in which they multiply before leaving the snail and encysting themselves on grass to begin the cycle again.

Affected sheep become very anaemic and lose condition. A watery swelling appears under the lower jaw, and the membranes of the eyes and mouth become very pale. If the presence of fluke is suspected, take a dung sample to your vet who will be able to tell you whether fluke eggs are present. The vet will also advise on the method of treatment most pertinent to the condition of your sheep. Cattle, goats, deer and rabbits can also act as hosts and carriers of fluke.

Roundworms

Roundworms occur in a variety of forms and those that cause problems for sheep start life as eggs passed out with the faeces onto the pasture. Given suitable climatic conditions of warmth and moisture, the eggs develop into larvae which go through three stages before migrating onto grass stems where they are eaten by sheep. Two weeks later they are mature adults, mating and producing eggs.

This is an oversimplification of the matter because different types of roundworm prefer different parts of the sheep's digestive tract in which to live and produce eggs. Some can lie dormant in the gut and hibernate until the condition of the sheep provides a springboard for a massive explosion of activity. Others hibernate through the winter as larval stages on the pasture, and only when the temperature rises in the spring (which also stimulates the growth of new grass) do they hatch in massive numbers to be eaten by young lambs and cause a lot of problems.

Diarrhoea, dehydration and reduced appetite are symptoms which can indicate worm infestations, although they can be confused with symptoms of other conditions. Even with light or moderate infestations there will be a reduced growth rate in lambs and the fleeces will appear dull, dry and lacking in bloom.

Anthelmintics are drugs that remove internal parasites and a vast range of them is available for the control of all kinds of roundworm. Clean pasture if available is invaluable in providing an efficient management aid in that newly dosed sheep can graze without the challenge of immediate re-infestation.

Tapeworms

Tapeworms are flat, ribbon-like parasites which live in the small intestine of the sheep.

The worm consists of a tiny head which attaches itself to the wall of the gut, either by a sucker or with hooks, and it feeds in that position. The ribbon-like body consists of a series of segments which are grown from the head and which may reach several metres in length. They are in fact packets of eggs which are shed from the end of the worm and passed out with the faeces. They require an intermediate host for their further development, in the same way as the sheep acts as an intermediate host for the tapeworm of the dog. Anthelmintics can be used to rid the sheep of this parasite, but unless the head is removed it will become a new worm and grow fresh segments. In spite of its size the tapeworm does very little damage to its host.

Coccidiosis

Coccidiosis is a disease of young lambs. Older sheep will have developed an immunity through previous contact with the parasite, but they can still pass on the infection. It occurs mostly in housed flocks although it can also infect lambs on heavily stocked pasture through contaminated faeces.

Infected lambs become unthrifty with a loss of appetite and develop a mild diarrhoea which causes dehydration. If coccidiosis is suspected, consult your vet over treatment and preventative measures to be taken in subsequent years.

Abortion

There are a number of different organisms which can cause abortion in sheep. Abortion can also be caused by other factors such as worrying dogs, malnutrition, incorrect handling and being forced too quickly through narrow gateways, to name but a few. In some cases of abortion early or midway through pregnancy the foetus may be so small that it passes unnoticed by the shepherd. A single *ewe* aborting may give no cause for alarm and be dismissed as one of the accepted risks of keeping livestock.

When a number of abortions occur, producing dead or very weakly foetuses, the vet should be called in without delay as it is only by microscopic bacteriological examination of aborted material that an accurate diagnosis can be made, and, if appropriate, the type of vaccine required to minimise further losses be recommended.

Domestic cats are responsible for spreading the infection causing one type of abortion in sheep, which is also contagious to humans. Consequently, extreme care should be taken when handling and disposing of any aborted material, and pregnant women should stay away from ewes at lambing time.

Navel Infections

The umbilical cord of a newborn lamb is moist and immediately after birth it is in contact with many different types of bacteria on the ground. This is the route by which certain harmful organisms gain entry to the body of the lamb. It is comparatively easy to prevent this from happening by applying an antiseptic dressing to the navel as soon as the lamb is born and then placing the lamb on a bed of clean straw. The aim is to dry up the cord as quickly as possible and to disinfect the area of the body around it.

Several preparations are suitable for use as a dressing:
- Tincture of iodine
- Antibiotic aerosol
- A weak solution of copper sulphate

A dressing which can be contained in a wide-necked bottle is preferable to an aerosol, as the wet cord and the area immediately surrounding it can be totally immersed in the liquid. Hold the lamb vertically by its forelegs and put the navel cord into the neck of the bottle. Lift the bottle at a right angle to the lamb and press it gently into the belly. (See illustrations on page 50.)

Blowfly Strike

From April to November the blowfly actively seeks sites where it can lay its eggs and the hatching maggots can obtain food. The fly finds the sheep particularly attractive, especially in the spring when young grass causes scouring and the liquefied faeces stick to the wool in the tail area. Even without the attraction of the faeces, the natural smell of the sheep is sufficient to entice the fly to lay its eggs on the wool and any open wounds suffer the same fate. Depending on the temperature, the eggs hatch in a few hours and the maggots start feeding by burrowing into the skin or open wounds of their host. The excreta from the maggots as they feed attracts more flies who lay more eggs, and in a short time unless treated, the body of the sheep will become a seething mass.

Long before this happens the behaviour of the sheep will indicate to the shepherd that it requires attention. Sheep which have newly hatched maggots will be irritated by them and will waggle their tails vigorously. They may stamp the ground with a hind foot and twist the whole body from side to side while holding the head high in the air. The sheep should be caught and examined carefully. Apply a combined insecticide and antiseptic to the site of the maggots and also the surrounding wool to catch any that attempt to escape, rubbing the medication into the fleece with the fingertips.

If the strike is left untreated the next indication will be a dark stain on the surface of the wool which is caused by the excreta from the maggots. By now the sheep will be feeling very sick, and may be lying by itself under a tree or hedge. If moved it will stand with its head lowered and quickly lie down again.

In July and August the blowfly quite

often lays its eggs on the back of the sheep near the shoulders. When the maggots develop there the sheep stands with its head lowered and appears to be very sick. The tail waggle and the body twisting are absent, and one of the few visual signs is the telltale dark stain starting at the back and spreading down to the shoulders. This appears after the maggots have been feeding for a day or two. Try to recognise and destroy them before this.

Any sheep which have bad footrot are also at risk. The smell that is so characteristic of footrot also attracts the fly to lay its eggs, and the resulting maggots penetrate the under-run sole of the foot. Trim the foot and saturate with insecticidal and antiseptic cream or liquid dressing, both of which are obtainable at your animal health store.

Adult sheep are at risk of fly strike until they are shorn. They are then free from infestation until the wool has grown to around 20mm, when a few weeks' protection may be had by dipping in an insecticidal dip. The fleece of the sheep being dipped will hold sufficient of the active ingredients of the insecticide to prevent the eggs from hatching. Depending on the weather and the frequency of the fly strikes, the lambs can be dipped separately at about the time the ewes are shorn. Any type of container which will hold sufficient dip to saturate the hindquarters and back will do, such as an old bath into which they can be slid on their backs and held for a few seconds before being lifted out.

Sheep Scab

Sheep scab is a contagious disease caused by a mite feeding on the skin of the sheep. Affected sheep rub against anything near at hand or scratch themselves with their feet. The activity of the mite produces scabs on the skin, and these, together with the rubbing, lead to the fleece falling out. Any type of skin irritation associated with rubbing should be viewed with suspicion. Sheep scab is a notifiable disease, and it should be reported at once to the Ministry of Agriculture, Fisheries and Food or the police.

At present there is a statutory regulation requiring *all* sheep to be dipped at a designated time of year for at least one minute in a dip approved by MAFF as effective for the control of sheep scab. This must be done in a swimbath in which the sheep can be completely immersed at least once. Up-to-date details of MAFF's requirements for its sheep scab eradication scheme are available from a local office.

To avoid the considerable expense of

Sheep dipping

installing a swimbath and its associated draining and holding pens, it might be possible to take the sheep to a nearby sheep farm and use the facilities there. Dipping is particularly stressful for sheep and they should not be dipped when full of food, thirsty or wet. Sheep that have been transported any distance should be rested before dipping and great care taken when handling pregnant ewes.

Several different compounds are approved for the control of sheep scab, and of these the organo-phosphorous type will also give protection from blowfly strike. It is also effective in reducing the tick population if the sheep are kept in a tick area, and will thus reduce the incidence of tickborne disease.

Problems can arise over the disposal of the dip solution left over after the dipping is completed. It is very toxic and should not be permitted to enter ponds, lakes, streams or ditches. Prior arrangements should be made with a waste disposal contractor who will be able to dispose of the dip solution safely and in compliance with the law. The National Rivers Authority will be able to advise you if you have any difficulties over disposal. In the interests of safety, the swimbath should be covered at all times when not being used for dipping sheep.

NOTIFIABLE DISEASES

In addition to sheep scab there are other important diseases which can affect sheep and when suspected they must be reported by the owner or person in charge of the animals to the police, or a veterinary officer of the MAFF in England, or the Department of Agriculture in Scotland. These are called notifiable diseases and a full list can be obtained from Animal Health Division A, Hook Rise South, Tolworth, Surbiton, Surrey KT6 7NF.

The list includes anthrax, foot and mouth disease, rabies and sheep pox which, together with sheep scab, are the diseases most likely to affect sheep.

ZOONOSES

Animal diseases which can affect human beings are called zoonoses, and while some may cause only slight problems, others can be dangerous and even life-threatening. Anyone who is working with sheep or in contact with them when a zoonotic disease is present or suspected should seek medical advice without delay if they feel unwell.

A list of zoonoses includes the following:
- anthrax
- rabies
- salmonellosis
- orf
- brucellosis
- chlamydiosis
- listeriosis
- toxoplasmosis
- all abortion agents

The last five on this list all pose a threat to pregnant women as they can cause abortion or serious defects if in contact with the unborn foetus. Pregnant women should keep away from sheep at lambing time.

The disposal of the carcases of any animals which die can be a problem. If it is impracticable to bury the carcase at home, the vet will advise on other methods of disposal which may be available to the flock owner.

As a postscript to this chapter on diseases in sheep, it should be pointed out that a great deal of money can be saved on call-out fees if you are able to take a sick sheep to your vet rather than have the vet visit you. If this is impracticable, emergency call-out fees can be saved by asking for a visit during the vet's normal working hours.

CHAPTER 9   **Keeping Records**

The small flock owner will find it extremely useful to keep a record of day-to-day events with the flock. The records can contain as little or as much of the history of the flock as the owner decides, and can include details relating to individual sheep, food purchased, grassland management, veterinary treatment, etc. Records of the performance of individual sheep start with accurate identification.

TATTOO

Identification

Each sheep in the flock should carry a permanent mark which will identify it and distinguish it from its flockmates in all the various forms of recording. In addition to an individual number, the mark can also indicate the year of birth and in some cases the sire.

Permanent identification can be one or more of several different kinds. It may consist of a tattoo mark in the ear, a metal or plastic tag in the ear, or a system of notches, indicating numbers, made around the outer edge of the ears with special pliers. Sheep with horns can be heat-branded, but it must be remembered that horns can be lost through accidents and fighting and the brand can become difficult to read in older sheep.

Tattoos are easier to read in the ears of white-faced sheep as they show up clearly when black tattooing ink is used. For sheep with coloured faces white or green tattooing ink is more suitable, but in all cases it is essential that the mark is placed centrally in the ear, between the two main sinews. With some pedigree breeds one ear is reserved for

EAR TAG

the flock number and the official registered mark of the breed society, while the other is left for the use of the flock owner.

Eartags come in a variety of shapes and sizes in both metal and plastic. Some metal tags are tamperproof in that they cannot be removed from the ear and re-used. The

An ear tag

type of plastic tag requires special pliers for application.

Ear notching, again using special pliers for the purpose, can be done at any age. It does not, however, indicate the year of birth, and for this a single letter tattooed in the ear is needed. The sheep's left ear is used for units and the right ear for tens. Five positions on the left ear will give sufficient spaces from one to nine, and from ten to ninety on the right. Sheep number seventy-nine would require notches denoting fifty and twenty on the right ear and five and four on the left, with a tattooed letter in the right ear indicating the year of birth.

EAR NOTCHING

manufacturers of these tags engrave numbers on them which are recorded to the name of the purchaser, although other numbers and letters can be engraved at the purchaser's request. The manufacturer's instructions should be carefully followed when using these tags on young lambs, remembering that the ear will grow and require more space within the tag. Special pliers are needed for their application. An advantage of using plastic tags is that they come in a variety of colours which can be seen at a distance. If a different colour is used for each year of birth, it is a great help in identifying a sheep without having to catch it. The tags also come in a variety of shapes and degrees of flexibility, but as they are all at risk from snagging on fences and equipment and can also on occasions break, it is advisable to put one in each ear. Each

Another method of identification is to use a temporary fleece brand. If this is to remain legible as the fleece grows, the size of the numbers should be fairly large. A metal rod of quarter-inch diameter is ideal for forming the numbers, which should measure 150mm form top to bottom. Short pieces of rod can be welded together to make up the number, and a handle welded on the back. These brands are used by dipping them into a shallow container of marking fluid, and positioning the number high up on the side of the sheep, rather than on the back. This will enable the number to be read from a distance, which is particularly useful at

FLEECE BRANDING

fleece. It is available in several different colours. The use of ordinary paint or tar will be heavily penalised by the British Wool Marketing Board when the fleece is graded, and the value of the whole fleece considerably reduced. This is because it is impossible to scour out these materials when the fleece is being processed.

Systems of recording

Once there is a system of identification, which is essential in pedigree flocks, records can be kept of the health and performance of the sheep.

Some breed societies have designed a pro-forma with headings under which details can be entered as they occur, and from which an accurate assessment can be made of the performance of the flock from year to year.

A loose-leaf system, based on one page for each sheep, can, for example, include details of the breed, the ear number, the age and a record of the sire and dam. In the case of a ewe, details of mating, the ram used, the date of service, together with the date of lambing and the sex and birth weight of the lambs born, any subsequent weighings, how disposed of and the price obtained can be recorded on a form like the one below.

For a ram, in addition to individual details, the number of ewes served, the total of lambs born to them, alive and dead,

lambing time when several people may be involved in looking after the sheep. It also avoids the necessity for catching individual sheep in order to identify them. The fleece brand need not necessarily be the same number as the ear number as long as a cross-reference list is kept. Thus the flock can be branded with consecutive numbers disregarding the individual numbers which will possibly contain a year letter and a lengthy figure.

Although curiously referred to as 'paint', the marking fluid that should be used is of a type approved by the British Wool Marketing Board as it can be scoured from the shorn

EWE RECORD CARD

BREED:		EAR No:		BORN:		SIRE:		DAM:	
Year	Ram Used	Date Lambed	Ear Nos & Sex of Lambs	Birth Weight	28 Day Weight	Disposal	Date	Wt. Price	Destination

together with details of the total sum realised when the lambs are sold, will allow a comparison to be made between the performance of the rams used, and between the performance of the same ram in different years.

In addition to the performance of individuals within the flock, a record of the sheep food purchased and times in the year when feeding was considered necessary, can be recorded. Any cultivations required by the grazing land and fertiliser applied, together with capital costs incurred for housing, fencing and equipment may provide interesting (or frightening!) information and be used in calculating a profit and loss account.

A separate record of all medicines purchased and administered, including the dates, is a legal requirement. To assist in this a booklet entitled "Animal Health Record Book", which incorporates the Code of Practice for the safe use of veterinary medicines on farms, has been produced by the National Office of Animal Health. Copies of this booklet may be obtained from MAFF whose officers have the statutory right to inspect the records kept.

The Movement of Animals (Records) Order 1960

There is a statutory requirement for all persons in charge of sheep to make a record of all movements onto or from the farm. A book for keeping this record can be obtained from the Animal Health department of the local authority, normally the County Council. If there is a major breakdown in animal health, this record is vital for tracing animals which might have been contacts at the source of the breakdown. The record of movement of animals must now be kept available for inspection for ten years.

The Animal Health Act 1981 and the Transit of Animals Order, copies of which may be obtained from H M Stationery Office, 49 High Holborn, London WC1V 6HB, require that details of the transport of sheep be recorded within eighteen hours of completion of the journey. The time of loading and unloading and the time when the animals are fed and watered must be entered as they take place. This is in addition to the Movement of Animals (Records) Order 1960, shown below.

FORM OF RECORD

MOVEMENT OF ANIMALS (RECORDS) ORDER 1960

Name and full address of person keeping the record ....

Date of Movement	Particulars of each bovine animal or deer moved to or from premises mentioned above				Number of sheep, goats or pigs (Specifying which)	Movements to premises mentioned above	Movements from premises mentioned above
	Breed	Age	Sex	Ear Mark or Ear Tag No.		Premises from which moved (including market, saleyard or fair) and/or Name and address of person from whom delivery was taken. N.B. – Both of those particulars are to be entered if available	Premises to which moved (including market, saleyard or fair) and/or Name and address of person taking delivery N.B. – Both of these particulars are to be entered if available
1	2	3	4	5	6	7	8

CHAPTER 10 The Shepherd's Calendar

In keeping and caring for sheep in flocks large or small, there follows a sequence of events throughout the year. It is the same for both early- and late-lambing flocks and begins in late summer or early autumn with the selection of breeding stock. The future crop of lambs will depend to a certain extent on daily observations by the shepherd of the prevailing conditions. The state of the grazing, the weather, supplementary food and preventive medicine all need to be assessed constantly and adjusted as necessary.

July, August, September

Make a thorough inspection of the ewe flock, paying particular attention to teeth, feet and udders. Mark for culling any that do not meet basic requirements, segregate them and sell them for meat.

If additional sheep are required, buy the best that can be afforded.

October

Get the ewes onto a rising plane of nutrition by putting them onto clean, fresh grazing or by feeding a small amount (100g) of concentrates each morning. Aim for a condition score of 3.5 at mating.

Fit the ram's harness in position a few days before mating is to begin and tighten if necessary when adding the coloured crayon.

Introduce the ram five months before the planned date for lambing, and record the date on which each ewe is served. The average gestation period is 147 days.

November

Stop feeding concentrate to the ewes when they have been mated, but feed hay if the quality and quantity of the grazing deteriorates. The hay must be of good quality, free from mould, and sweet-smelling. It should be fed from covered hayracks as sheep will not eat wet hay.

December, January, February

Consider housing the ewes if the weather is bad, using barley straw for bedding. Continue feeding hay ad lib. Introduce concentrates six weeks before lambing is due to begin. Vaccinate each ewe with a booster dose of vaccine against clostridial diseases immediately before lambing.

Prepare the lambing kit and ensure that all records are up to date.

March, April

Lambing. Record lamb dates of birth and birth weights. Carry out tailing and castrating.

Continue feeding ewes, adding succulent foods such as roots, cabbage type green food or soaked sugar beet pulp if possible.

Creep feed lambs.

Give ewes a dose of broad spectrum anthelmintics and trim feet. Weather permitting, turn ewes and lambs out onto clean grazing by day, returning to a communal pen in the sheep house at night. After a period of hardening off they can be left out overnight.

Dose lambs against worms at six weeks of age and thereafter at monthly intervals.

Foot trimming using a sheep stretcher

May, June

Reduce the concentrate feeding for the ewes as the quantity and quality of the grazing increases. Offer hay in covered hayracks if there is any scouring.

Carry out crutching, shearing and dipping against fly strike.

July

Wean lambs onto fresh, clean grazing. Sell lambs destined for the meat market as they reach 40kg.

Vaccinate any remaining lambs against clostridial diseases as they reach sixteen weeks of age.

August, September

Attend store lamb, breeding ewe and ram sales. Make up the breeding flock.

CHAPTER 11 **Showing**

Many small flock owners get pleasure from taking their sheep to shows and competing for the honour of winning prize cards and rosettes. Many local shows now have classes for all types of sheep and lambs, and as they are usually held at weekends it means that more people can spare the time to attend.

All sheep shown, whether pure or crossbred, should have good conformation and be well-grown for their age. Depending on the breed, they should be in good condition, with the down breeds having a condition score of at least 3.5 or 4. The fleece should be clean, the feet well trimmed and it is a great help for both sheep and shepherd if the sheep is taught to walk on a halter. Halter trained sheep are able to show themselves off to the judge without involving the handler in a rodeo routine!

Selection

When selecting a sheep for show, start looking well before the date of the show and select an animal that has the indefinable quality of 'character'. This is the sheep that is always appealing to the eye in any position and one that always stands out from its flock-mates. If it is a purebred or pedigree sheep, check the list of points which have been drawn up by the breed society concerned as being particularly desirable. Match the selected sheep as nearly as possible to that list. Select a sheep with a bold eye and an alert expression, a straight back and standing four square on straight legs. Observe the way it moves when walking towards and away from you, especially when being led on a halter.

Halter training

Some show judges ask for an entry to be turned loose in the show ring so that they can see for themselves how the sheep moves naturally, but others ask for it to be paraded on a halter. Use a cotton halter for training rather than one made from polypropylene and fit a piece of rubber tubing to the part of the halter which goes under the lower jaw. This will prevent the

A SHEEP HALTER

75

Teaching to lead on a long rope

halter from becoming too tight if the sheep puts up a lot of resistance, and thus prevent chafing the skin which might lead to an unsightly scar. Have a long lead rope on the halter when beginning the training and walk behind the sheep in such a way that it can be restrained if it tries to break away. A push from behind will usually encourage it to walk forward. After a half-hour session each day for a few days, it will quickly become accustomed to the halter and the handler, and walk beside him in a natural way.

The next step is to get the animal to stand still while on the halter and to stand in a way in which it shows itself off to the best advantage. It may be necessary to move a leg slightly to achieve the best position. This can be done either by gently pushing or pulling the sheep to one side or the other, or by moving it with the hand.

Standing correctly

A mixed breed class

Showmanship

Wear a clean white coat and keep the sheep in a correct position all the time that it is before the judge. It should not be allowed to graze or turn around or lie down. Onlookers might be amused to see this, but it could cost a prize if it happened just when the judge looked back in order to make a comparison with another sheep. Keep the head of the sheep at a naturally comfortable height while it is being held, neither too high nor too low. Stand in front of the sheep while the judge is standing behind the line-up, but move to one side so that he can see the front when he comes around to the other side. Watch the judge as well as the sheep all the time and try to convince him by showmanship and without words that you have the best sheep!

A champion ram

Dressing

The fleeces carried by the various breeds of sheep vary a great deal between the fine, short staple of the Southdown and the long, coarse wool of the Scottish Blackface. Others like the Wensleydale have long, lustrous fleeces which hang in ringlets, and the Herdwick and Welsh Mountain contain a high proportion of kemp in their wool.

Different types of wool require different treatments before being presented for the judge's inspection. Tradition dictates that on some breeds unnatural shapes are created, using hand shears, while others have a minimum of trimming with a programme of dipping in a variety of dips. In all cases, the fleece should be clean and free from dung and vegetable matter.

Any washing that is required should be done about ten days before the show. Tie up the sheep and soak with water from a hosepipe or buckets. Use soap flakes and work up a lather in the wool, using the fingers to tease out any matted locks. If a swimbath is available, fill it with clean water and allow the sheep to swim in it to rinse out the soap and dirt. Keep the sheep on a halter while swimming, as fat sheep with heavy fleeces can be at risk in the bath. Use the halter to pull the sheep to the ramp leading out of the bath if the animal appears to be distressed in any way. If no swimbath is available for rinsing, use a hosepipe or bucket with plenty of clean water.

Choose a dry day for the washing as it is important that the sheep dries quickly. Never put the sheep under cover while still wet. Washing with soap will leave the fleece soft, white and open in texture, and it will respond to being dipped in a bath containing show dip. These dips are designed specifically to firm up the wool in shortwools and to add to the lustre and purl effect in longwools. They also take away the flat whiteness which is apparent after washing in soap.

As each breed of sheep has its own traditional shape or appearance when brought out for show, it is not proposed here to go into the requirements for each breed. Suffice it to say that if the plan is to produce show entries trimmed with hand shears, then take a course by an expert on how it should be done. The equipment required — hand shears, sharpening stone, carding wire on heavy wooden backs, halters and an old shearing comb for cleaning the carding wire — is modest, but the skill in using it is not!

Take consolation from the fact that a good sheep is attractive to the experienced eye whether it is trimmed or not.

Lambs can be shown in their natural fleece, while older sheep will look better if they are carrying about four months' growth of wool. All that is required on a sheep that has been docked is that the wool on the end of the tail is squared off parallel to the ground.

The Texel breed has stipulated that all its show animals must be shown in a natural condition. Any trimming or shaping is not allowed, although faces and legs can be washed clean.

Fleece wool

A less demanding type of showing is in classes for shorn fleeces. Select the fleeces while they are still on the sheep and swim the sheep in clean water without soap, dip or disinfectant added, around four weeks before shearing. Ask the shearer to take special care when shearing and wrap the shorn fleece according to the instructions in the show schedule. Place the fleece in a large paper sack as this will help it to retain its wrapped shape. Avoid plastic sacks which will cause the wool to sweat. Identify your fleeces with a clean well-written label.

A rolled fleece

Showing butcher's lambs

To show lambs in the butcher's lamb class requires a knowledge of carcase classification. This is a way by which the conformation and degree of fatness is described by visual appraisal, and given a code by the Meat and Livestock Commission staff working in the abattoir. The conformation classification is graded on the scale E,U,R,O,P with E being the best and P the poorest. The fatness is assessed by handling the tail and the spinous and transverse processes of the loin and grading between 1 (the leanest) and 5 (the fattest).

The preferred weight of the carcase is between fifteen and twenty kilograms, so the aim is for a finished carcase grading of 2 or 3 in the conformation class E or U. Grades 3, 4 and 5 are further subdivided into L and H, with L denoting low fat cover and H denoting high. The preferred classification by most meat buyers is within the range of 2 and 3L. The weight of the carcase will be around 45 per cent of the liveweight of the lamb, depending on the degree of fatness.

Great care should be taken at all times with loading and transporting lambs, but particularly if the lamb is to be shown as a carcase. It should never be held or picked up by the wool as this will cause distress to the lamb and result in unsightly bruising of the carcase. Avoid any sharp projections which could cause injury in the transport used to take the lambs to the show, and also avoid abscesses on the carcase caused by faulty injection techniques. Injection sites should be high up on the neck and the hypodermic needles used should be sharp and sterile.

Butcher's lambs

CHAPTER 12 **Grassland Management**

The flock owner with only a small area of land on which to graze the sheep will perhaps find it difficult to do very much in the way of constructive management to bring about improvement in the grass. Nitrogenous fertiliser used in the early springtime will give a boost to the growth of grass but it must be balanced by adding phosphate and potash at some time in the year or the pasture will become impoverished.

Chain harrowing when the pasture is dry in the early spring and before housed sheep are turned out with their lambs, will remove dead material from the turf and scatter droppings. It will also encourage new growth in the grass.

The pasture can be improved by the clearance of any unpalatable weeds, nettles, thistles, docks, etc. If the acreage is small, the docks can be dug out by hand and the nettles and thistles, if cut regularly through the growing period, will eventually succumb. The sheep will often eat the cut material once it has wilted. Ragwort can pose a problem. The sheep will graze it when it is a rosette of leaves in the early spring without any ill-effects, but later in the season if allowed to grow to the flowering stage it is poisonous to sheep. If it gets included in hay or is cut and left to wilt and then eaten, the accumulation of poison will eventually kill the animal.

Regular mowing to contain the surplus grass even while the sheep are still grazing will improve the pasture. Once the grass gets to the flowering stage, it does not produce much more leaf, which is the nutritious part of the plant. Regular mowing will prevent the grass from flowering or running to seed, thus promoting further growth of leaf. Regular mowing will also prevent weeds from seeding and increasing their population in the sward. More of the sheep droppings will be exposed to sunlight which will lead to a lower rate of infestation by internal parasites because of the effect it has on the eggs of the roundworms.

If it is possible to subdivide the pasture more involved procedures can be undertaken. The old turf can be treated with a herbicide to kill it off and a rotovator used to cultivate the surface before sowing new grass and clover seed. When choosing new grass seed, select leafy strains of ryegrass for the main part of the mixture and add white clover and deep-rooting herbs. White clover with its ability to fix nitrogen from the air, to the benefit of the accompanying grass, is also high in energy, protein, minerals and vitamins. In addition to giving variety to the mixture, the herbs chicory, burnet and yarrow will help by bringing up minerals from below the level to which the roots of the grass penetrate.

Any quantity of surplus grass can be fenced off using temporary electric fencing and be made into hay. In view of the machinery involved, this is a job for a local contractor. Cut the grass when it is at the early flowering stage and if possible get it dry and baled without allowing any rain on it, then stack it under cover. The regrowth, or aftermath, provides valuable feed for newly weaned lambs as it is clean and the risk of parasitic infection is much less, due to the break in the life cycle of the parasites when the hay crop is taken off.

Subdividing the pasture

CHAPTER 13

Miscellaneous Information Concerning Sheep

SOME PLANTS POISONOUS TO SHEEP

Most poisonous plants are avoided by sheep unless grazing is very sparse as in times of drought or during spells of severe weather, especially snowstorms. The main danger comes when they are cut, wilted and made into hay, often unnoticed, and are then readily eaten by the sheep.

Yew   Probably the plant causing most problems for the small flock owner is the yew. If only a small quantity is eaten the animal will die. This applies to the living shrub or tree, and also to any hedge trimmings which may inadvertently be dumped where the sheep have access to them.

Bracken is poisonous to stock but to a much lesser extent than yew, and will normally be eaten only when grazing is in short supply or when it is made into hay or used as bedding. Sheep affected by bracken poisoning become blind.

Equisetum or horsetail is not usually a problem unless cut, dried and made into hay. If it is then eaten, it is poisonous and there is no antidote. The sheep dies.

Laburnum   Sheep seem to be less affected by eating this tree than some other types of livestock. All parts of the tree are poisonous and will cause death if consumed in quantity.

Acorns are sometimes eaten in quantity by sheep who inexplicably develop a craving for them. They are more poisonous when green and unripe. Affected animals suffer from constipation, which is then quickly followed by diarrhoea. They should be moved away from the source as quicky as possible.

Azaleas and rhododendrons often 'escape' into boundary fences from gardens and woodlands and can cause considerable damage to the nervous system if eaten. As with yew, the clippings from gardens are as toxic when wilted as when they are green. The shrub should either be removed completely or fenced off out of reach of the sheep, as ewes seeking extra nutrition in the weeks just prior to lambing will readily eat it, having avoided it completely for months before.

Ragwort should be pulled out and burned before it reaches the flowering stage (see Chapter 12). The toxin accumulates in sheep if they are given hay containing the dried plant.

SHEEP NAMES

In different parts of the country, different names are given to the various ages of sheep and this can be confusing for the new flock owner. Some names are in common usage, however, and lamb is the universal name given to the newly born. The lambs can be either ram lambs or ewe lambs and they carry this name through the summer even after they are weaned. When the lamb wool breaks to form locks, the wool then forms a tapering point on each lock, similar to the tag on an old-fashioned shoe lace, and tag, or more commonly teg, is the word used to describe this age of sheep. The term hogg or hogget is also used in some parts of the country.

A ewe and her lambs are spoken of as a couple and where there are twins as double couples.

Chilver is the name given to females in south-west England, and gimmer is commonly used in Scotland and the north of England to describe female sheep after their first shearing. Theave is also a name used in parts of England for once-shorn ewes, while in all areas shearling refers to both males and females between their first and second shearings. Subsequently they can be known as two-shear, three-shear or four-shear rams or ewes.

Sometimes the number of permanent incisor teeth present is used to describe the age of the sheep, and a shearling becomes a two-tooth, a two-shear a four-tooth, a three-shear is a six-tooth and a four-shear is known as full-mouthed.

In hill flocks a substantial income is derived from selling cast ewes. These are ewes which have reached maturity and still have sound teeth and udders, and which can be replaced by the annual intake of shearling ewes bred in the flock. The cast ewes are sold to lowland farmers for crossbreeding where they do not have to withstand the rigours of the climate often experienced by hill sheep. They are sometimes described as draft ewes in other parts of the country. A broken-mouthed ewe or crone is one which has lost some incisor teeth through age but which may still be capable of breeding and rearing lambs, provided that the udder is sound.

A ewe which does not breed when given the opportunity to do so is called either barren, guest or eild.

A cull ewe is one which is not retained for breeding for any reason — chronic ill-health, poor conformation, or as a non-breeder — and which is sold with no guarantee as to its fitness for future breeding.

A wether is a castrated male sheep, and a rig is a male sheep which has either one or both testicles undescended.

SOME ORGANISATIONS SERVING THE SHEEP INDUSTRY

The National Sheep Association
  The Sheep Centre, Malvern, Worcs
  WR13 6PH
  (0684) 892661
The Rare Breeds Survival Trust
  National Agricultural Centre, Kenilworth,
  Warks CV8 2LG
  (0203) 696551
Organic Sheep Society
  Worcester College of Agriculture,
  Hindlip, Worcs WR3 8SS
  (0905) 513103
British Wool Marketing Board
  Oak Mills, Station Road, Clayton,
  Bradford, West Yorkshire BD14 6JD
  (0274) 882091
National Agricultural Centre
  Stoneleigh, Kenilworth, Warks CV8 2LG
  (0203) 686503
Meat & Livestock Commission
  P.O. Box 44, Winterhill House, Milton
  Keynes, Bucks MX6 2DQ
  (0908) 677577
Royal Agricultural Society of England
  National Agricultural Centre, Stoneleigh,
  Kenilworth, Warks CV8 2LZ
  (0203) 696969

Advice on the design of sheep housing may be obtained from ADAS at the local offices of the Ministry of Agriculture, Fisheries and Food. The address can be found in the local telephone directory.

MAKING EQUIPMENT

It is possible to save considerable amounts of money by making certain pieces of equipment rather than buying them. An example of this is a hay rack.

Use softwood as sawn. Two 150mm × 25mm planks for lower part of sides, and one 150mm × 25mm for the upper part, nailed or screwed to 75mm × 50mm × 850mm corner posts, with a gap of 350mm between upper and lower parts. Total length not more than 2.5m. Side planks projecting 50mm beyond corner posts at each end,

and nailed or screwed to a central upright of 75mm × 50mm × 850mm.

Ends 850mm long with planks nailed to corner posts and finishing flush with them. Four 12mm metal tie rods connecting sides to ends through holes drilled through side planks and corner posts, and secured with nuts and washers.

Use 100mm × 100mm weld mesh cut slightly smaller than the inside area of the hay rack placed on the hay to prevent wastage, or make a 'ladder' from 50mm × 25mm for the same purpose.

If used outside, cover with corrugated iron.

12mm diam tie rod   150mm × 25mm   12mm diam tie rod

75mm × 50mm   200   350 mm   75mm × 50mm   300 mm

150mm × 25mm

← 2.5m →   ← 850mm →

(Not drawn to scale)

HAY RACK

84

TRANSPORT

There are occasions when it is necessary to transport members of the flock. It might be to a show, to a market or fair or to a dipping or shearing centre. There is a statutory obligation under the Animal Health Act 1981 and the Transit of Animals (Road and Rail) Order 1975 as amended for certain legal requirements to be observed when transporting stock.

The main ones apply to all forms of transport and may be summarised under the following headings:

Ventilation should be adequate and appropriate to the number of stock being transported.

The interior of the transport vehicle shall be free from projections likely to cause injury or suffering.

The floor shall be of non-slip design and/or strewn with straw or similar material to provide a proper foothold.

Ramps shall be designed to prevent slipping and have no gap between the ramp and the floor of the transport vehicle in which a foot or leg can be trapped.

Sidegates are to be of sufficient height and strength to avoid falls.

Partitions/dividers must be fitted if the floor exceeds 3.1m in length. If a full load is not carried, the pens will need to be smaller.

A sheep trailer made by one of the authors

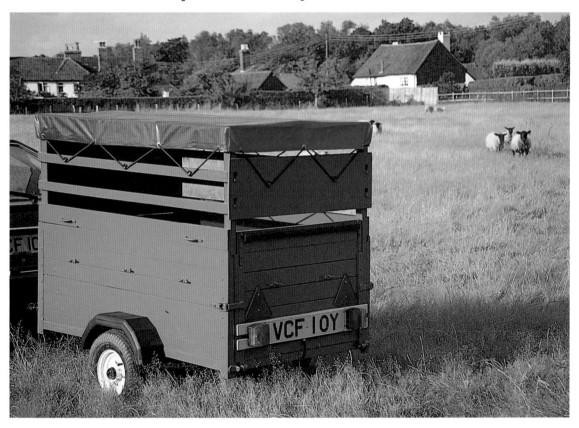

SOME SUPPLIERS OF
SHEEP EQUIPMENT

Alfred Cox (Surgical) Ltd.
 Edward Road, Coulsden, Surrey
 CR5 2XA
 (081) 6682131
 A wide range of sheep breeding
 equipment, veterinary instruments, and
 hand tools used in caring for sheep.

TWIL Group Marketing Ltd. Fencing
Manufacturers
 P.O. Box 119, Shepcote Lane, Sheffield
 S9 1TY
 Stock fencing. Free copy of Sentinel
 Fencing Handbook.

Dalton Supplies Ltd.
 Nettlebed, Henley-on-Thames, Oxon
 RG9 5AB
 (0491) 641457
 Ear tags and identification systems and
 livestock equipment.

Gallagher Agricultural Ltd.
 Curriers Close, Canley, Coventry
 CV4 8AW
 (0203) 470141
 Portable and permanent electric fencing.
 Free planning service for electric fencing
 systems.

Richey Tagg Ltd.
 Fearby Road, Masham, Ripon,
 N. Yorks. HG4 4ES
 (0765) 89541
 Permanent and temporary sheep
 identification, anti-fighting masks for
 rams, lambing equipment and general
 sheep care products.

Shearwell Ltd.
 Putham Farm, Wheddon Cross,
 Minehead, Somerset TA24 4AS
 (0643) 841552
 Sheep requisites, sheds, handling and
 feeding systems, slatted flooring. Quality
 sheep also available.

IN CONCLUSION

"Sheep are seldom boring and rarely profitable!" to quote a well-known farmer. In spite of this (or because of it?) some twenty million of the creatures are kept in this country. Endearing and infuriating all at the same time, they provide a magnetic interest for those who allow themselves to be attracted to them.

Treat them well and the rewards will be many.

Happy sheep keeping!

Index

For a free illustrated book list ring (0473) 241122 or write

**Farming Press Books, 4 Friars Courtyard
30-32 Princes Street, Ipswich IP1 1RJ, United Kingdom**